Lecture Notes in Mathematics

Edited by A. Dold, Heidelberg and B. Eckmann, Zürich

382

Jacobus H. van Lint

University of Technology Eindhoven, Eindhoven/Netherlands

Combinatorial Theory Seminar Eindhoven University of Technology

Springer-Verlag

Berlin · Heidelberg · New York 1974

AMS Subject Classifications (1970): 05 A 15, 05 A 17, 05 B 10, 05 B 15, 05 B 20, 05 B 25

ISBN 3-540-06735-3 Springer-Verlag Berlin · Heidelberg · New York
ISBN 0-387-06735-3 Springer-Verlag New York · Heidelberg · Berlin

© by Springer-Verlag Berlin · Heidelberg 1974. Library of Congress Catalog Card Number 74-2551. Printed in Germany.

Offsetdruck: Julius Beltz, Hemsbach/Bergstr.

PREFACE

These Lecture Notes are the work-out of a seminar held at the Technological University Eindhoven (THE) in the years 1971-1972 and 1972-1973. As a guide for the seminar the book "Combinatorial Theory" by Marshall Hall, Jr. was chosen. Since this book is used by so many combinatorialists it was considered worthwhile to publish our notes as a service to the mathematical community. The contents fall into the following categories: anwers to questions which came up during the seminar, extensions and generalizations of theorems in Hall's book, references and reports on results which appeared after the book, and finally a number of research results of members of the group.
The members of the seminar were M.L.J. Hautus, H.J.L. Kamps, J.H. van Lint, K.A. Post, C.P.J. Schnabel, J.J. Seidel, H.C.A. van Tilborg, J.H. Timmermans and J.A.P.M. van de Wiel. The author of these notes acted as leader of the seminar. A number of valuable suggestions is due to N.G. de Bruijn.
The chapters in these notes have the same titles as those in Hall's book and the notation is the same. References to this book are preceded by H., e.g. H. Theorem 8.3.2 or (H.8.3.10); definitions and theorems are not repeated.
For her excellent typing of these lecture notes I thank Mrs. E. Baselmans-Weijers.

J.H. van Lint.

Eindhoven, November 1973

CONTENTS

I. PERMUTATIONS AND COMBINATIONS

Since there is extensive literature on the subject of permutations, combinations and on the use of binomial coefficients (e.g. [1], [2]), we restrict ourselves to some comments on this chapter.

1.1. *Combinations of* n *things taken* r *at a time, etc.*

The result (H.1.1.4) can be derived in a different way. As an example we represent the combination a a b f from the set {a,b,c,d,e,f} by aa|b|-|-|-|f or alternatively by (2,1,0,0,0,1). Clearly we have $n-1$ marks and r letters. The position of the $n-1$ marks completely determines the combination. Hence the required number is $\binom{r+n-1}{n-1}$. The representation (2,1,0,0,0,1) makes it clear that this is the number of solutions of $x_1 + x_2 + \ldots + x_n = r$ in nonnegative integers. In connection with this result and (H.1.1.5) and (H.1.1.6) it is worthwhile to remark that one can also count the number of solutions in nonnegative integers x_i of

$$x_1 + x_2 + \ldots + x_n \le r \tag{1.1.1}$$

by using a 1 to 1 mapping, namely,

$$(x_1, x_2, \ldots, x_n) \mapsto (x_1 + 1,\ x_1 + x_2 + 2,\ \ldots,\ x_1 + x_2 + \ldots + x_n + n)\ . \tag{1.1.2}$$

This yields a combination without repetition from $\{1,2,\ldots,n+r\}$.
It follows that the number of solutions of (1.1.1) is $\binom{n+r}{n}$. In a sense this is a more combinatorial way to solve the problem than the obvious one of summing (H.1.1.4) as follows:

$$\sum_{k=0}^{r} \binom{n+k-1}{k} = \binom{n+r}{n}\ ,$$

a well known relation for binomial coefficients.
The answer to (H.1.1.7) can also be derived as follows. Take any combination $x_1 < x_2 < \ldots < x_n$ from the integers $1,2,\ldots,n-r+1$ and consider the mapping $(x_1, x_2, \ldots, x_r) \mapsto (x_1, x_2 + 1, \ldots, x_r + r - 1)$ which yields a combination from $\{1,2,\ldots,n\}$ with no repeats and no consecutive integers. The number $\binom{n-r+1}{r}$ is now immediately clear. This result generalizes to circles in the following way. If we are to choose r points from n on a circle, with no two consecutive, we first number the points on the circle $1,2,\ldots,n$. Now distinguish between r-tuples including 1 and those not including 1. Clearly by the previous result the number of such allowable r-tuples is respectively $\binom{n-r-1}{r-1}$ and $\binom{n-r}{r}$. Adding these numbers we find that there are

$$\frac{n}{n-r}\binom{n-r}{r} \tag{1.1.3}$$

r-tuples from n on a circle, with no two consecutive.

The following argument is simpler. Choose one point on the circle (n possibilities). Leave out this point and its neighbors and apply the linear result to $r-1$ chosen from $n-3$. Every solution has now been obtained r times. Hence the required number is

$$(1.1.4) \qquad \frac{n}{r}\binom{n-r-1}{r-1} = \frac{n}{n-r}\binom{n-r}{r}$$

(cf. (H.2.1.24)).

We shall return to this problem in (2.3).

1.2. *Identities involving binomial coefficients.*

The formulas in (H.1.1.9) are not correct. For $n = 0$ the sum (H.1.1.9b) is 1, for $n = 1$ the sum (H.1.1.9c) is -1 and finally for $n = r$ the sum (H.1.1.9d) is $(-1)^r$. We derive (H.1.1.9d) in a different way. Note that

$$\sum_{k=r}^{\infty} (-1)^k \binom{k}{r} x^k = (-1)^r x^r (1+x)^{-r-1} .$$

It follows that

$$(1.2.1) \qquad \sum_{k=r}^{n} (-1)^k \binom{k}{r}\binom{n}{n-k}$$

is the coefficient of x^n in the product $(-x)^r(1+x)^{-r-1}(1+x)^n$. If $r = n$ we have $(-x)^n(1+x)^{-1} = (-1)^n x^n + \ldots$ and if $n > r$ then $(-x)^r(1+x)^{n-r-1}$ is a polynomial of degree $n-1$ and therefore the sum in (1.2.1) is 0.

The formula for the multinomial coefficient (H.1.1.10) can also be derived by induction on r. If we denote by $F_r(b_1,b_2,\ldots,b_r)$ the number to be determined and we consider the possible positions of the b_r objects of the r-th kind we find

$$F_r(b_1,b_2,\ldots,b_r) = \binom{b_1+b_2+\ldots+b_r}{b_r} F_{r-1}(b_1,b_2,\ldots,b_{r-1})$$

from which (H.1.1.10) immediately follows. The advantage of this method is that it also is applicable in the following more difficult problem, which occurs in the 400 "best problems" from the American Mathematical Monthly 1918-1950 (cf. [3], # 3731).

PROBLEM. In how many ways can a_1 1's, a_2 2's, ..., a_n n's be arranged, so that in reading from the beginning, none of the $(k+1)$'s are reached until at least one of the k's has been reached $(k = 1,\ldots,n-1)$.

First solution. If any a_i is 0, then the number is 0, so we may assume all a_i to be positive. Let $F_n(a_1,a_2,\ldots,a_n)$ be the required number. Clearly the first number is a 1 and the remaining 1's can be placed arbitrarily. If we leave out the 1's we find a sequence of the same type using only 2,3,...,n. Hence

$$F_n(a_1,a_2,\dots,a_n) = \binom{a_1+a_2+\dots+a_n-1}{a_1-1} F_{n-1}(a_2,\dots,a_n) \; .$$

From this we find

$$F_n(a_1,a_2,\dots,a_n) = \prod_{i=1}^{n} \binom{a_i+a_{i+1}+\dots+a_n-1}{a_i-1} = \tag{1.2.2}$$

$$= \frac{(a_1+a_2+\dots+a_n)!}{a_1!\,a_2!\dots a_n!} \prod_{i=1}^{n} \frac{a_i}{a_i+a_{i+1}+\dots+a_n}$$

in which the first factor is the familiar multinomial coefficient.

Second solution. Now that we have the answer (1.2.2) to our problem, it is not difficult to find a more combinatorial argument. Consider all permutations of a_1 1's, a_2 2's, ..., a_n n's. The fraction of these which have a 1 in the first position is $\frac{a_1}{a_1+a_2+\dots+a_n}$. If we consider only these permutations and then strike out all the ones, then we have a set of permutations of 2's,3's,...,n's. Of these, the fraction $\frac{a_2}{a_2+a_3+\dots+a_n}$ has a 2 in the first position. Now (1.2.2) immediately follows by induction.

References:

[1] J. Riordan, An Introduction to Combinatorial Analysis, John Wiley & Sons, New York, 1958.

[2] J. Riordan, Combinatorial Identities, John Wiley & Sons, New York, 1968.

[3] The Otto Dunkel Memorial Problem Book, Am. Math. Monthly 64 (1957).

II. INVERSION FORMULAE

2.1. *The principle of inclusion and exclusion; Permanents.*

In this chapter we wish to make use of permanents. For this reason we present the generalized sieve formulae as treated by Ryser [1]. The principle of inclusion and exclusion is a special case of this formula. Let S be a set of n elements and let F be a field (in fact, we only use that F is a group). To each element a of S we assign a *weight* $w(a) \in F$. If $P_1, P_2, \ldots, P_N$ are subsets of S then we define

$$W(P_{i_1}, P_{i_2}, \ldots, P_{i_r}) := \sum_{a \in P_{i_1} \cap P_{i_2} \cap \ldots \cap P_{i_r}} w(a) ,$$

$$W(r) := \sum_{1 \le i_1 < i_2 < \ldots < i_r \le N} W(P_{i_1}, P_{i_2}, \ldots, P_{i_r}) \qquad (r \ge 1) ,$$

$$W(0) := \sum_{a \in S} w(a) .$$

Then we have

THEOREM 2.1.1. *If* E(m) *is the sum of the weights of the elements of* S *which are in exactly* m *of the subsets* P_i $(i = 1,2,\ldots,N)$, *then*

$$E(m) = \sum_{i=0}^{N-m} (-1)^i \binom{m+i}{m} W(m+i) . \qquad (2.1.1)$$

DEFINITION. *Let* $A := [a_{ij}]$ *be an* m *by* n *matrix* $(m \le n)$. *Then* per(A), *the permanent of* A, *is defined by*

$$\operatorname{per}(A) := \sum_{(i_1, i_2, \ldots, i_m)} a_{1i_1} a_{2i_2} \cdots a_{mi_m}$$

where $(i_1, i_2, \ldots, i_m)$ *runs through all permutations without repetition of* m *elements from* $\{1,2,\ldots,n\}$.

The following properties of the permanent are fairly obvious:

(2.1.2) (i) If $m = n$ then $\operatorname{per}(A) = \operatorname{per}(A^T)$,

(2.1.3) (ii) If P and Q are permutation matrices then $\operatorname{per}(PAQ) = \operatorname{per} A$,

(2.1.4) (iii) A permanent can be calculated by expansion in the same way as a determinant. The permanent is a linear function of each of the row-(column)-vectors which form A.

The following method of calculating a permanent is an application of Theorem 2.1.1, given by Ryser [1]. If A is given, we form A_r by leaving out r columns of A and then we calculate the product of the row sums of A_r, which we call $S(A_r)$. Fixing r and choosing A_r in all possible ways, we find $\Sigma_r := \Sigma\, S(A_r)$. Then

$$\text{(2.1.5)} \qquad \operatorname{per}(A) = \sum_{i=0}^{m-1} (-1)^i \binom{n-m+i}{i} \Sigma_{n-m+i} .$$

The proof is an easy application of (1.1) (cf. [1]).

We now consider a problem concerning binomial coefficients (cf. Chapter I, reference [3], problem 3625).

PROBLEM. Prove the relation

$$\text{(2.1.6)} \qquad \sum_{r=0}^{n-1} (-1)^r \binom{n}{r} (n-r)^n = n! .$$

First solution. Let J be the n by n matrix of 1's. Then clearly per(A) = n!. In the notation used above we have $\Sigma_r = \binom{n}{r}(n-r)^n$. Hence (2.1.6) follows from (2.1.5).

Second solution. Define

$$\text{(2.1.7)} \qquad f_n(x) := \sum_{k=0}^{\infty} k^n x^k \qquad (n = 0,1,\ldots) .$$

By differentiating (2.1.7) we find

$$\text{(2.1.8)} \qquad x f_n'(x) = f_{n+1}(x) .$$

Since $f_0(x) = (1-x)^{-1}$ it follows from (2.1.8) that

$$f_n(x) = (1-x)^{-n-1} P_n(x) ,$$

where $P_0(x) = 1$ and $P_n(x)$ is a polynomial of degree n satisfying the relation

$$\text{(2.1.9)} \qquad P_{n+1}(x) = x(1-x)P_n'(x) + (n+1)x\, P_n(x) .$$

Taking x = 1 in (2.1.9) we find $P_{n+1}(1) = (n+1)P_n(1)$, i.e. $P_n(1) = n!$. The sum on the left hand side of (2.1.6) is the coefficient of x^n in the product $(1-x)^n f_n(x)$, i.e. the coefficient of x^n in $(1-x)^{-1} P_n(x)$, which is $P_n(1)$. This proves (2.1.6).

We shall return to this problem in Chapter III (cf. (3.2.9)).

2.2. *Derangements.*

The number of derangements of $1,2,\dots,n$ (i.e. permutations fixing no element) is generally calculated by using the principle of inclusion and exclusion as was done for (H.2.1.7). We demonstrate a different inversion method.

THEOREM 2.2.1. *Let* $f(n)$ *and* $g(n)$ *be functions defined for nonnegative integers, satisfying*

$$f(n) = \sum_{i=0}^{n} \binom{n}{i} g(n-i) = \sum_{i=0}^{n} \binom{n}{i} g(i) . \tag{2.2.1}$$

Then we may invert this relation to express g *in terms of* f *by the rule*

$$g(n) = \sum_{i=0}^{n} (-1)^i \binom{n}{i} f(n-i) . \tag{2.2.2}$$

Proof. A straightforward proof can be given by substitution, namely

$$\sum_{i=0}^{n} (-1)^i \binom{n}{i} f(n-i) = \sum_{i=0}^{n} (-1)^i \binom{n}{i} \sum_{j=0}^{n-i} \binom{n-i}{j} g(n-i-j) =$$

$$= \sum_{m=0}^{n} \binom{n}{m} g(n-m) \sum_{i=0}^{m} (-1)^i \binom{m}{i} = g(n) ,$$

since the inner sum is 0 for $m > 0$.

A proof which gives more insight in the meaning of (2.2.1) is obtained by considering the (formal) power series

$$F(x) := \sum_{n=0}^{\infty} f(n) \frac{x^n}{n!} ,$$

$$G(x) := \sum_{n=0}^{\infty} g(n) \frac{x^n}{n!} .$$

If we also consider the sums in (2.2.1) and (2.2.2) as coefficients of power series, then Theorem 2.2.1 says that $F(x) = e^x G(x)$ implies $G(x) = e^{-x} F(x)$.
We apply Theorem 2.2.1 to obtain the number of derangements of $1,2,\dots,n$. Note that $\binom{n}{r} D_{n-r}$ is the number of permutations $a_1,a_2,\dots,a_n$ of $1,2,\dots,n$ such that $a_i = i$ for exactly r values of i. Hence

$$\sum_{r=0}^{n} \binom{n}{r} D_{n-r} = n! . \tag{2.2.3}$$

By Theorem 2.2.1 we find from (2.2.3) that

(2.2.4) $$D_n = \sum_{i=0}^{n} (-1)^i \binom{n}{i}(n-i)! = n! \sum_{i=0}^{n} \frac{(-1)^i}{i!} .$$

We remark that by the second proof of the theorem

(2.2.5) $$\sum_{n=0}^{\infty} D_n \frac{x^n}{n!} = e^{-x} \sum_{n=0}^{\infty} x^n = e^{-x}(1-x)^{-1} .$$

Let us now consider E_n, the number of *even* permutations of $1,2,\dots,n$ which are derangements. This number can be calculated in exactly the same way as (H.2.1.6). Remark that the number of even permutations fixing r specified numbers is equal to half of the number of all permutations fixing these r numbers for $r \le n-2$, and exactly equal to the latter for $r = n-1$ and $r = n$. Hence

(2.2.6) $$E_n = \tfrac{1}{2}n! - \tfrac{1}{2}\binom{n}{1}(n-1)! + \dots + \tfrac{1}{2}(-1)^{n-2}\binom{n}{n-2}2! + (-1)^{n-1}\binom{n}{n-1}1! + (-1)^n\binom{n}{n}0! =$$

$$= \tfrac{1}{2}D_n + \tfrac{1}{2}(-1)^{n-1}(n-1)$$

(cf. Am. Math. Monthly 79 (1972), Problem E2354).

2.3. *Ménage numbers.*

We consider once again the problem of choosing r points from n on a circle, with no two adjacent. We show a connection with (H.2.1.21) and (H.2.1.24) in the case $(n,r) = 1$. In this case periodicity is not possible and the required number is therefore equal to n times the number of circular combinations of r from $n-r$. Since $(n-r,r) = 1$ we find from (H.2.1.21) with $b_1 := r$, $b_2 := n-2r$, that this number is $\frac{1}{n-r}\binom{n-r}{r}$ from which (1.1.3) immediately follows. The restriction $(n,r) = 1$ can now be dropped. A configuration on the circle with n positions which has period n/k is mapped into a configuration on the circle with $n-r$ positions which has period $(n-r)/k$. Hence, instead of the factor $\frac{n}{n-r}$ we now find $\frac{n/k}{(n-r)/k}$.

Consider the n by n matrix A defined by

(2.3.1) $$a_{ij} := \begin{cases} 0 & \text{if } j-i \equiv 0 \text{ or } 1 \pmod{n} , \\ 1 & \text{else.} \end{cases}$$

By the definition we see that the ménage numbers U_n (cf. (H.2.1.25)) satisfy

(2.3.2) $$U_n = \mathrm{per}(A_n) .$$

Now we define the n by n matrix B_n by

(2.3.3) $$b_{ij} := \begin{cases} 0 & \text{if } j-i = 0 \text{ or } 1 , \\ 1 & \text{else.} \end{cases}$$

From (2.1.4) we find

$$\text{(2.3.4)} \qquad \operatorname{per}(B_n) = \operatorname{per}(A_n) + \operatorname{per}(B_{n-1}) ,$$

which is a relation between the ménage problem for a circular table and the same problem for a "linear table", i.e. the number of permutations of 1,2,...,n such that $\forall_i\ [a_i \notin \{i,i+1\}]$.

In order to establish a recurrence for the permanent of B_n we introduce the notation $B_n^{(i,j)}$ for the matrix formed by replacing a 0 in position (i,j) of B_n by a 1. By convention $B_n^{(n,n+1)}$ shall mean B_n. By expanding by the first row we find

$$\text{(2.3.5)} \qquad \operatorname{per} B_{n+1} = \sum_{i=2}^{n} \{\operatorname{per} B_n^{(i-1,i)} + \operatorname{per} B_{n-1}^{(i-1,i)}\} =$$

$$= \sum_{i=2}^{n} \{\operatorname{per} B_n^{(i,i)} + \operatorname{per} B_{n-1}^{(i-1,i-1)}\} .$$

From (2.1.4) we find by row and column operations

$$\operatorname{per}\begin{pmatrix} B_k^T & J_{k,\ell} \\ J_{\ell,k} & B_\ell \end{pmatrix} = \operatorname{per} B_{k+\ell} ,$$

where $J_{p,q}$ is the p by q matrix of ones. We use this several times in the following expansions of permanents. If we apply (2.1.4) to (2.3.5) we find:

$$\text{(2.3.6)} \qquad \operatorname{per} B_n^{(1,1)} = \operatorname{per} B_n^{(n,n)} = \operatorname{per} B_n + \operatorname{per} B_{n-1} ,$$

$$\text{(2.3.7)} \qquad \operatorname{per} B_n^{(i,i+1)} = \operatorname{per} B_n + \operatorname{per} B_{n-1}^{(i,i)} ,$$

and

$$\text{(2.3.7a)} \qquad \operatorname{per} B_n^{(i,i)} = \operatorname{per} B_n + \operatorname{per} B_{n-1}^{(i-1,i)} ,$$

and a number of similar relations. Substitution of (2.3.6) and (2.3.7) in (2.3.5) leads to the relation

$$\text{(2.3.8)} \qquad \operatorname{per} B_{n+1} = n\{\operatorname{per} B_n + \operatorname{per} B_{n-1}\} + \operatorname{per} B_{n-2} .$$

We shall now find a number of expressions for per B_n and the ménage numbers U_n, following a method suggested by N.G. de Bruijn. We shall choose a suitable contour K in the complex plane and we then try to find a function ψ such that

$$\text{(2.3.9)} \qquad \operatorname{per} B_n = \int_K \psi(p) p^n \, dp .$$

The contour K is chosen in such a way that integration by parts of $\psi(p)(p+1)p^n$ yields a first term 0. Each of the three contours K_i in figure 1 is a candidate.

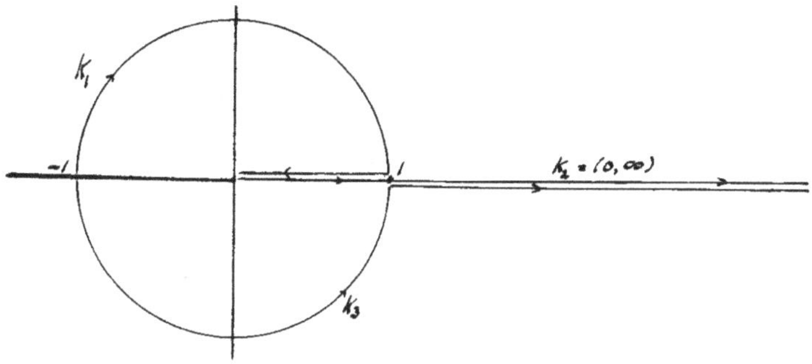

Fig. 1

Substituting (2.3.9) in (2.3.8) we find

(2.3.10) $$\psi'(p) = \{-1 + \frac{1}{p^2(p+1)}\}\psi(p) .$$

The solution of (2.3.10) is

(2.3.11) $$\psi(p) = (1 + p^{-1})e^{-(p+p^{-1})} .$$

Each of the K_i and (2.3.11) substituted in (2.3.9) yields a solution of the recurrence (2.3.8). If we extend (2.3.8) to negative integers we find

(2.3.12) $$\text{per } B_n = - \text{per } B_{-n-1}$$

which should be interpreted formally only.
The 3 choices for K in (2.3.9) give us 3 functions of n with different behavior for $n \to \infty$, resp. $n \to -\infty$. Hence these solutions are linearly independent.
The solution of (2.3.8) which we need is a linear combination of the 3 functions, i.e.

(2.3.13) $$\text{per } B_n = \sum_{i=1}^{3} c_i \int_{K_i} e^{-(p+p^{-1})}(p+1)p^{n-1}\,dp .$$

Such a combination satisfies (2.3.12) only if $c_1 = c_3$ and $c_2 = 0$.
Now let K be $K_1 + K_3$. Then

(2.3.14) $$\text{per } B_n = c \int_{K} e^{-(p+p^{-1})}(p+1)p^{n-1}\,dp .$$

To find c we substitute $n = -1$, use $e^{i\theta}$ resp. $e^{-i\theta}$ as parameters for the semicircles, and transform (0,1) into $(1,\infty)$. We find

$$\text{per } B_{-1} = -1 = c\{-2\int_0^{\pi} e^{-2\cos\theta}\sin\theta\, d\theta - \int_1^{\infty} e^{-(x+x^{-1})}\, d(x+x^{-1})\} =$$

$$= -c\int_{-2}^{\infty} e^{-u}\, du = -c\,e^2 ,$$

hence $c = e^{-2}$. Thus we have proved

$$\text{per } B_n = e^{-2}\int_K e^{-(p+p^{-1})}(p+1)p^{n-1}\, dp . \tag{2.3.15}$$

Let C_n be the n-th Chebyshev polynomial of the second kind, i.e.

$$C_n(x) := 2T_n(\tfrac{x}{2}) \quad \text{and} \quad T_n(\cos\theta) = \cos n\theta$$

where T_n is the Chebyshev polynomial of the first kind (cf. [2] Ch. 22).
We now consider the ménage numbers U_n. By (2.3.2), (2.3.4), (2.3.12) and (2.3.15) we have

$$U_n = \tfrac{1}{2}e^{-2}\int_K e^{-(p+p^{-1})}(p^{n-2} + p^{-n-2})(p^2 - 1)dp = \tag{2.3.16}$$

$$= e^{-2}\int_{-1}^{\infty} e^{-(x+x^{-1})}(x^n + x^{-n})(1 - x^{-2})dx .$$

Using the relation $C_n(x + x^{-1}) = x^n + x^{-n}$ (cf. [2] Ch. 22) we find from (2.3.16)

$$U_n = e^{-2}\int_{-2}^{\infty} e^{-u}\, C_n(u)du = \int_0^{\infty} e^{-y}\, C_n(y-2)dy . \tag{2.3.17}$$

The explicit expression for $C_n(x)$ is

$$C_n(x) = n\sum_{m=0}^{[n/2]} (-1)^m \frac{(n-m-1)!}{m!(n-2m)!}\, x^{n-2m} , \tag{2.3.18}$$

(cf. [2] Ch. 22). We remark that (2.3.17) and (2.3.18) imply

$$U_n \sim e^{-2}\, n! \qquad (n \to \infty) \tag{2.3.19}$$

(cf. H.Ch. 2, Problem 3).
We shall now show that (H.2.1.25) follows from (2.3.17) in two ways. First we substitute (2.3.18) in (2.3.17):

$$U_n = n \sum_{m=0}^{[n/2]} (-1)^m \frac{(n-m-1)!}{m!(n-2m)!} \sum_{j=0}^{n-2m} \binom{n-2m}{j} j!(-2)^{n-2m-j}$$

from which (H.2.1.25) follows by changing the order of summation and then calculating the inner sum. An easier method is to use $x^n + x^{-n} = C_{2n}(x^{\frac{1}{2}} + x^{-\frac{1}{2}})$. Then (2.3.16) yields

$$U_n = \int_0^\infty e^{-u} C_{2n}(u^{\frac{1}{2}})du , \tag{2.3.20}$$

an expression which is derived in a completely different way in [3], Problem 18 on p.228. Now substitution of (2.3.18) in (2.3.20) yields

$$U_n = \sum_{m=0}^{n} (-1)^m \frac{2n}{2n-m} \binom{2n-m}{m} (n-m)! , \tag{2.3.21}$$

which is (H.2.1.25).
(For another solution using permanents cf. [6] p.21.)

2.4. *Incidence algebras and Möbius functions.*

We wish to take a closer look at case 2 on H.p.15. We make the usual convention for arithmetical functions f that $f(x) := 0$ if $x \notin \mathbb{N}$. Then the zeta function ζ of the incidence algebra $A(\mathbb{N})$ satisfies

$$\zeta(m,n) = f_1(\tfrac{n}{m}) ,$$

where f_1 is the arithmetical function which is identically 1.
If $g_1(m,n) := \mu(\frac{n}{m})$ we find by (H.2.2.1) and H. Lemma 2.1.1

$$(\zeta g_1)(m,n) = \sum_{d,\, m|d,\, d|n} \zeta(m,d) g_1(d,n) = \sum_{d' | \frac{n}{m}} \mu(\tfrac{n/m}{d'}) =$$

$$= \begin{cases} 1 & \text{if } m = n, \\ 0 & \text{if } m \neq n. \end{cases}$$

Hence the Möbius function $\mu(m,n)$ of $A(\mathbb{N})$ is given by $\mu(m,n) = \mu(\frac{n}{m})$.
Now consider the subalgebra $A^*(\mathbb{N})$ of $A(\mathbb{N})$ consisting of all functions $f \in A(\mathbb{N})$ for which there is an arithmetical function F such that $f(m,n) = F(\frac{n}{m})$. Note that if f and g are two such functions and $h = fg$, then

$$h(m,n) = \sum_{d,\, m|d,\, d|n} f(m,d) g(d,n) = \sum_{d' | \frac{n}{m}} F(d') G(\tfrac{n/m}{d'}) ,$$

which is a function of $\frac{n}{m}$, i.e. there is a function H such that $h(m,n) = H(\frac{n}{m})$. This shows that $A^*(\mathbb{N})$ is indeed a subalgebra. For this subalgebra H. Theorem 2.2.1 is simply the classical Möbius inversion formula for arithmetical functions (H. Theorem 2.1.1). To place everything in the proper setting we associate with each $f \in A^*(\mathbb{N})$ a formal Dirichlet series $\hat{F}(s) := \sum_{n=1}^{\infty} F(n)n^{-s}$. If $fg = h$ then we have

$$\hat{F}(s)\hat{G}(s) = \Big(\sum_{n=1}^{\infty} F(n)n^{-s}\Big)\Big(\sum_{m=1}^{\infty} G(m)m^{-s}\Big) = \sum_{\ell=1}^{\infty} \ell^{-s}\Big(\sum_{\ell|n} F(n)G(\tfrac{\ell}{n})\Big) = \hat{H}(s) .$$

Hence $A^*(\mathbb{N})$ is mapped isomorphically in the algebra of formal Dirichlet series with the usual sum and product operation. It is this setting which explains the use of the symbol ζ because we now have

$$\hat{\zeta}(s) = \sum_{n=1}^{\infty} n^{-s} = \zeta(s) ,$$

the Riemann zeta function.

Another amusing example of Möbius inversion is obtained if we take $\mathbb{N}$ with the usual ordering. In this case we find from $\delta(m,n) = \sum_{m\le d\le n} \mu(d,n)$ that

$$\mu(m,n) = \begin{cases} 1 & \text{if } m = n , \\ -1 & \text{if } m = n-1 , \\ 0 & \text{otherwise.} \end{cases} \tag{2.4.1}$$

In this case the Möbius inversion formula says that, if

$$g(n) = \sum_{k=1}^{n} f(k) \quad \text{then} \quad f(n) = g(n) - g(n-1) ,$$

the trivial relation between the terms and the partial sums of a series!

As the next example let $P := \{x \in \mathbb{R} \mid x \ge 1\}$ and introduce the ordering

$$x \preccurlyeq y \quad \text{iff} \quad yx^{-1} \in \mathbb{N} .$$

Now the Möbius inversion formula says that if

$$g(x) = \sum_{y \preccurlyeq x} f(y) = \sum_{n\le x} f(\tfrac{x}{n}) \tag{2.4.2}$$

then

$$f(x) = \sum_{y \preccurlyeq x} g(y)\mu(y,x) = \sum_{n\le x} \mu(n)g(\tfrac{x}{n}) , \tag{2.4.3}$$

since again $\mu(y,x) = \mu(\frac{x}{y})$ (cf. [4] Theorem 268).

A well known result on the Möbius function is obtained by applying this last inversion formula to the function f which is identically 1. By (2.4.2) we have $g(x) = [x] = x + O(1)$ and now (2.4.3) yields

$$\sum_{n\le x} \mu(n)[\tfrac{x}{n}] = 1 ,$$

hence

$$x \sum_{n\le x} n^{-1} \mu(n) + O([x]) = 1 ,$$

or

$$(2.4.4) \qquad \sum_{n\le x} n^{-1} \mu(n) = O(1) .$$

We give one example of H. Theorem 2.2.1 without the zero element, in which the sums are not finite. Introduce an ordering on $\mathbb{N}$ by $m \preccurlyeq n$ iff $mn^{-1} \in \mathbb{N}$.
Let $\sum_{n=1}^{\infty} f(n)$ be an absolutely convergent series. If we define

$$g(n) := \sum_{k=1}^{\infty} f(kn) = \sum_{m\preccurlyeq n} f(m) ,$$

then (H.2.2.7) again holds. In fact we have

$$\sum_{m\preccurlyeq n} \mu(m,n)g(m) = \sum_{k=1}^{\infty} \mu(k)g(kn) = \sum_{k=1}^{\infty} \mu(k) \sum_{\ell=1}^{\infty} f(\ell kn) =$$

$$= \sum_{r=1}^{\infty} f(rn) \sum_{k|r} \mu(k) = f(n) .$$

2.5. *An application of Möbius inversion.*

We consider mappings $F: (GF(2))^m \to GF(2)$. Each such mapping is representable by a polynomial in m variables $x_i \in GF(2)$, of degree ≤ 1 in each of the variables. If F is such a polynomial we define

$\nu(F)$:= the number of variables x_i which do not occur in F,

$|F|$:= the number of monomials of F.

E.g., if $m = 5$ and $F(x_1,\dots,x_5) = x_1x_2 + x_3x_4$ then $\nu(F) = 1$, $|F| = 2$. We shall write $G \subset F$ if every monomial of G is also a monomial of F. For a fixed polynomial F we consider the set of all monomials of F. We order the subsets of this set as in H.p.15 case 1. Each of the subsets corresponds to a polynomial $G \subset F$. We define the function f on the subsets by

(2.5.1) $f(G)$:= the number of m-tuples $(x_1,x_2,\dots,x_m)$ for which every monomial of G has the value 0 and every other monomial of F has the value 1.

It is then easily seen that

$$\sum_{H \subseteq G} f(H) = 2^{\nu(F-G)} . \tag{2.5.2}$$

Using (H.2.2.10) we apply Möbius inversion to (2.5.2):

$$f(G) = \sum_{H \subset G} (-1)^{|G|-|H|} \, 2^{\nu(F-H)} . \tag{2.5.3}$$

This inversion formula enables us to prove a theorem on the number of zeros of polynomials over GF(2). The theorem is useful a.o. in coding theory (cf. [5], Ch. 6).

THEOREM 2.5.1. *If* F *is a polynomial in* m *variables over* GF(2) *and* N(F) *is the number of zeros of* F *then*

$$N(F) = 2^{m-1} + \sum_{G \subset F} (-1)^{|G|} \, 2^{|G|+\nu(G)-1} .$$

Proof. By (2.5.1) we have

$$N(F) = \sum_{G \subset F, \, |F-G| \equiv 0 \pmod 2} f(G) .$$

Furthermore, it is clear that

$$\sum_{G \subset F} f(G) = 2^m .$$

Hence

$$\begin{aligned} N(F) &= 2^{m-1} + \tfrac{1}{2} \sum_{G \subset F} (-1)^{|F-G|} f(G) = \\ &= 2^{m-1} + \tfrac{1}{2} \sum_{G \subset F} (-1)^{|F-G|} \sum_{H \subset G} (-1)^{|G|-|H|} \, 2^{\nu(F-H)} = \\ &= 2^{m-1} + \tfrac{1}{2} \sum_{H \subset F} (-1)^{|F|-|H|} \, 2^{\nu(F-H)} \sum_{H \subset G \subset F} 1 = \\ &= 2^{m-1} + \tfrac{1}{2} \sum_{H \subset F} (-1)^{|F-H|} \, 2^{\nu(F-H)} \, 2^{|F-H|} , \end{aligned}$$

which proves the theorem.

For the example mentioned above, $m = 5$ and $F(x_1,\dots,x_5) = x_1x_2 + x_3x_4$, we have 4 polynomials G with $G \subset F$, namely 0, x_1x_2, x_3x_4 and F. We find

$$N(F) = 16 + 16 - 8 - 8 + 4 = 20 .$$

2.6. *Permutations with restricted position.*

At the time H.p.18, Problem 2 was discussed in the seminar, it was conjectured by H.J.L. Kamps that the number of permutations $(a_1,a_2,\ldots,a_{2n})$ of $1,2,\ldots,2n$, such that no column of the array

$$\begin{array}{l} 1,2,\ldots,\ n\ ,\ n+1,n+2,\ldots,2n \\ n,1,\ldots,n-1,\ \ 2n,n+1,\ldots,2n-1 \\ a_1,a_2,\ \ldots\ldots\ldots\ldots\ldots\ ,a_{2n} \end{array}$$

has a repeated number, is $U_{2n} + 2$ for $n \geq 2$. Here we remark that (H.2.1.26) is true for $n \geq 4$ and that the recurrence can be used for $n < 4$ to define U_n for $n \leq 1$. This leads to $U_1 = -1$, $U_0 = 2$. The conjecture was then generalized and proved in two ways, namely by a combinatorial argument and also by calculating permanents. We now state the theorem and then give both proofs.

THEOREM 2.6.1. *Let* U_n *denote a ménage number as usual and let* $U_{m,n}$ *be the number of permutations* $(a_1,a_2,\ldots,a_{m+n})$ *such that no column of the array*

$$\begin{array}{l} 1,2,\ldots,\ m\ ,\ m+1,m+2,\ldots,m+n \\ m,1,\ldots,m-1,\ m+n,m+1,\ldots,m+n-1 \\ a_1,a_2,\ \ldots\ldots\ldots\ldots\ldots\ ,a_{m+n} \end{array} \tag{2.6.1}$$

has a repeated number. Then for $m \geq n \geq 2$

$$U_{m,n} = U_{m+n} + U_{m-n}\ . \tag{2.6.2}$$

First Proof. The permutations as in (2.6.1) and those counted by the ménage numbers can be symbolized by the same figure (see fig. 2).

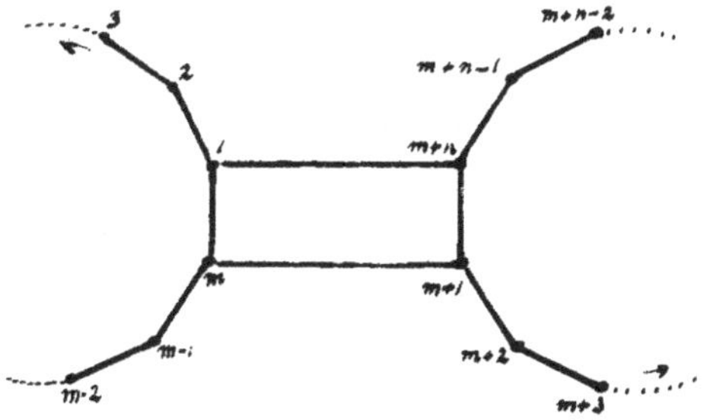

Fig. 2

Here we have points numbered $1,2,\ldots,m+n$ joined by edges as shown, where for U_{m+n} we include the horizontal edges $(1,m+n)$ and $(m,m+1)$ instead of the vertical edges $(1,m)$ and $(m+n,m+1)$ which we take in the case of $U_{m,n}$. Both problems amount to adjoining $1,2,\ldots,m+n$ to the edges in such a way that no edge is numbered with a number adjoined to one of its terminal vertices. Now we observe:

(i) If $\{x,y\} \cap \{1,m,m+1,m+n\} = \emptyset$ $(x \neq y)$, then the pair $\{x,y\}$ can be adjoined to the edges $(1,m)$ and $(m+n,m+1)$ in two ways and the same holds for the edges $(1,m+n)$ and $(m,m+1)$. The identity mapping on the remaining edges then provides us with a one-to-one mapping of permutations of the specified type in the two classes.

(ii) The pair $\{1,m+1\}$ can be adjoined to the edges $(1,m+n)$ and $(m,m+1)$ in one way. The same holds for the edges $(1,m)$ and $(m+n,m+1)$. Once again the identity mapping on the other edges maps permutations of these types into each other (one-to-one). We treat the pair $\{m,m+n\}$ analogously.

(iii) The pairs $\{1,m\}$ and $\{m+1,m+n\}$ can be adjoined to the edges $(1,m+n)$ and $(m,m+1)$ but this cannot be done for the edges $(1,m)$ and $(m+n,m+1)$. The converse is true for the pairs $\{1,m+n\}$ and $\{m,m+1\}$.

From (i), (ii), (iii) we see that our counting problem is reduced to comparing permutations counted by U_{m+n}, having the pair $\{1,m\}$ adjoined to edges $(1,m+n)$ and $(m,m+1)$ (we call these type I), with permutations counted by $U_{m,n}$, having the pair $\{1,m+n\}$ adjoined to edges $(1,m)$ and $(m+n,m+1)$ (we call these type II). Of course $\{m+1,m+n\}$ and $\{m,m+1\}$ are then treated in the same way.
Now consider the following similar problem. Consider a figure as in fig. 3.

Fig. 3

Here we have two "chains" of vertices joined by edges, where the terminal edges can end in a vertex or they can be open. Using the conventional notation for intervals we symbolize figure 3 by $[k]$, (ℓ). Note that in figure 3 there are $k + \ell$ edges. We define:

DEFINITION. *The symbol* $\mathcal{P}([k],(\ell))$ *denotes the set of permutations* $(a_1,a_2,\ldots,a_{k+\ell})$ *of* $(1,2,\ldots,k+\ell)$, *which, when adjoined sequentially to the edges of figure* 3, *have the property that no edge has a number equal to the number of one of its terminal vertices.*

Let us now return to figure 2 and consider the permutations which we called "type I" and "type II" above. For these permutations we leave out the edges $(1,m+n)$ and $(m,m+1)$, respectively $(1,m)$ and $(m+n,m+1)$ and in either case also leave out the vertices which have a number corresponding to an edge which has been dropped. Finally, we renumber the vertices etc. What has happened is that our type I permutations have become the set $\mathcal{P}((m-2),[n])$ and the type II permutations are now the set $\mathcal{P}((m-1],[n-1))$.
Since we wish to postpone the actual counting as long as possible we now try to find a natural one-to-one mapping from the set $\mathcal{P}([k],(\ell))$ to the set $\mathcal{P}([k-1),[\ell+1))$. For this purpose consider figure 4 in which a permutation $\pi := (a_1,a_2,\ldots,a_{k+\ell})$ has been

adjoined to the edges of the figures corresponding to the two classes

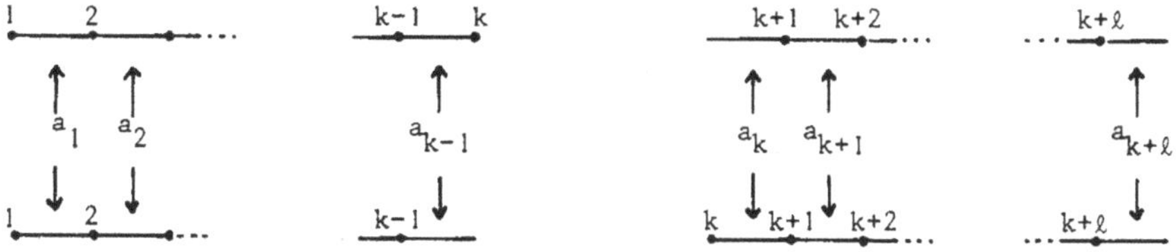

Fig. 4

Clearly, if $\pi \in \mathcal{P}([k],(\ell))$ then also $\pi \in \mathcal{P}([k-1),[\ell+1))$ unless $a_k = k$. Conversely, $\pi \in \mathcal{P}([k-1),[\ell+1))$ implies $\pi \in \mathcal{P}([k],(\ell))$ unless $a_{k-1} = k$. Hence we already have a natural mapping from one class to the other, namely the identity, except for the "exceptional" permutations mentioned above. So, only these remain to be considered. If $\pi \in \mathcal{P}([k],(\ell))$ has $a_k = k$ then we leave out vertex k and edge k. We do the same for a $\pi \in \mathcal{P}([k-1),[\ell+1))$ with $a_{k-1} = k$. Finally we renumber and we are then in the situation of figure 5.

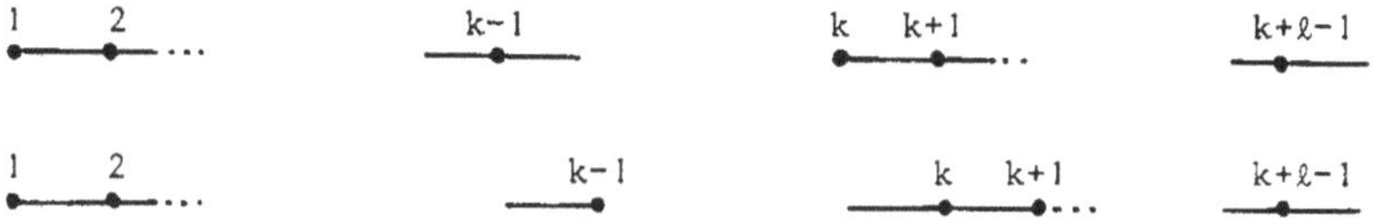

Fig. 5

Hence we must now map $\mathcal{P}([k-1),[\ell))$ into $\mathcal{P}([k-1],(\ell))$. Once again, for most permutations we can map with the identity. In the same way as above there are exceptional permutations which have to be considered separately. For these we again leave out a vertex and an edge and then we renumber. We then find ourselves with $\mathcal{P}([k-1],(\ell-1))$ and $\mathcal{P}([k-2),[\ell))$. This is the original problem with both parameters reduced by one. One step in the induction we have just described consists of two reductions. Now we distinguish a number of cases.

(a) Case $k = \ell$. After $k-1$ steps we have the situation of figure 6.

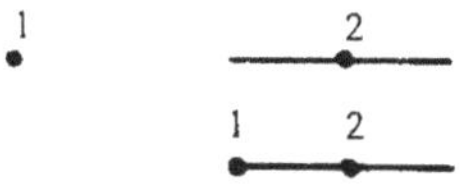

Fig. 6

The classes $\mathcal{P}([1],(1))$ and $\mathcal{P}([0),[2))$ are both empty. In this case we have found a one-to-one mapping of $\mathcal{P}([k],(\ell))$ onto $\mathcal{P}([k-1),[\ell+1))$.

(b) Case $k = \ell + 1$. The induction now ends with the situation of figure 7.

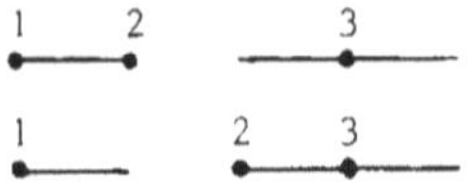

Fig. 7

Here $|\mathcal{P}([2],(1))| = 2$ and $|\mathcal{P}([1),[2))| = 1$, so we can complete the one-to-one mapping with the exception of exactly one permutation in $\mathcal{P}([k],(\ell))$.

(c) Case $k \geq \ell + 2$. After ℓ steps of the induction we are in the situation of figure 8.

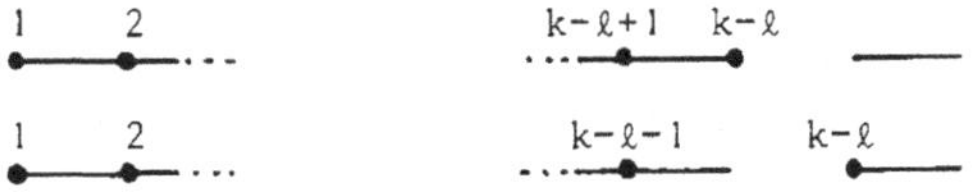

Fig. 8

We now do one more reduction, i.e. half of an induction step. Then the upper half no longer has exceptional permutations and the exceptional permutations for the lower half are exactly those counted by per $B_{k-\ell-2}$ by our definition (2.3.2). We have shown

(2.6.3) $$|\mathcal{P}([k],(\ell))| - |\mathcal{P}([k-1),[\ell+1))| = -\operatorname{per} B_{k-\ell-2} \,.$$

(d) Case $k = \ell - 1$. In the same way as above we find a one-to-one mapping.

(e) Case $k \leq \ell - 2$. Now the reduction leads to

(2.6.4) $$|\mathcal{P}([k],(\ell))| - |\mathcal{P}([k-1),[\ell+1))| = \operatorname{per} B_{\ell-k+1} \,.$$

The results we have found in cases (a) to (e) combined with (2.3.11) show that (2.6.4) holds for all positive values of k and ℓ.

Returning to our original problem we have now shown that

$$\begin{aligned} U_{m,n} - U_{m+n} &= -\operatorname{per} B_{m-n-1} - \operatorname{per} B_{n-m-1} = \\ &= \operatorname{per} B_{m-n} - \operatorname{per} B_{m-n-1} = \\ &= U_{m-n} \end{aligned}$$

by (2.3.2), (2.3.4) and (2.3.12).

This proves Theorem 2.6.1. The ideas for this proof were provided by N.G. de Bruijn. The second proof is a straightforward brute force attack using permanents.

Second Proof. Let $J_{p,q}$ be the matrix of 1's of size p by q. For the present counting problem we need the following generalizations of (2.3.1) and (2.3.3):

(2.6.5) $$A_{m,n} := \begin{pmatrix} A_m & J_{m,n} \\ J_{n,m} & A_n \end{pmatrix} ,$$

(2.6.6) $$B_{m,n} := \begin{pmatrix} B_m & J_{m,n} \\ J_{n,m} & B_n \end{pmatrix} .$$

Note, that by our definition preceding (2.3.5) we have

(2.6.7) $$B_{m,n} = B_{m+n}^{(m,m+1)} .$$

Now, $U_{m+n} = \text{per } A_{m+n}$ by (2.3.2) and

(2.6.8) $$U_{m,n} = \text{per } A_{m,n} .$$

The matrices $A_{m,n}$ and A_{m+n} differ only in the m-th row and the (m + n)-th row. If we develop the permanents by these rows and compare, we find

(2.6.9) $$\text{per } A_{m,n} = \text{per } A_{m+n} + 2 \text{ per } B_{m+n-2}^{(m-1,m)} - \text{per } B_{m+n-2}^{(m-1,m-1)} - \text{per } B_{m+n-2}^{(m,m)} .$$

We reduce the right hand side by applying both (2.3.7) and (2.3.7a). In the B-permanents this amounts to decreasing all upper indices by 1 and all lower indices by 2. The same terms occur if we apply (2.6.9) for m - 1 and n - 1. Hence it follows that

(2.6.10) $$\text{per } A_{m,n} - \text{per } A_{m+n} = \text{per } A_{m-1,n-1} - \text{per } A_{m+n-2} .$$

By (2.6.8) we find from (2.6.10)

(2.6.11) $$U_{m,n} - U_{m+n} = U_{m-1,n-1} - U_{m+n-2}$$

and then by induction and one more application of (2.6.8), (2.6.9), (2.3.6), (2.3.7) and (2.3.7a) we arrive at

$$U_{m,n} - U_{m+n} = U_{m-n}$$

which proves Theorem 2.6.1.

A lot of the work for this proof had already been done in section 2.3. This makes it look shorter than it actually is. The first proof, though longer, has the advantage that it is more a combinatorial proof. A closer look shows that the proofs are essentially the same!

For a general theory of permutations with restricted positions see [3].

References:

[1] H.J. Ryser, Combinatorial Mathematics, Carus Math. Monograph 14, 1963.

[2] Handbook of Mathematical Functions, Nat. Bureau of Standards Appl. Math. Ser. 55, 1964.

[3] J. Riordan, An Introduction to Combinatorial Analysis, John Wiley & Sons, New York, 1958.

[4] G.H. Hardy and E.M. Wright, An Introduction to the Theory of Numbers, Oxford Univ. Press, 1954.

[5] J.H. van Lint, Coding Theory, Lecture Notes in Math. 201, Springer Verlag, Berlin, 1971.

[6] J.K. Percus, Combinatorial Methods, Springer Verlag, New York, 1971.

III. GENERATING FUNCTIONS AND RECURSIONS

3.1. *The recursion* $u_n = \sum_{i=1}^{n-1} u_i u_{n-i}$.

In H. § 3.2 the combinatorial problem of counting the number of ways a sequence $x_1, x_2, \ldots, x_n$ may be combined in this order by a binary nonassociative product is treated. This leads in a natural way to the recursion in the title. The solution of the problem is

$$(3.1.1) \qquad u_n = \frac{1}{n}\binom{2n-2}{n-1}, \qquad n \geq 1 .$$

The same result, generally derived from the same recursion, is found for many other combinatorial problems. We shall list a number of these problems below and then give a number of combinatorial demonstrations that these problems indeed have the same solution. The sequence $(u_n)_{n\in\mathbb{N}}$ is known as the *Catalan* sequence. A bibliography of 243 papers and books in which the Catalan numbers occur can be found in [8].

PROBLEM 1. The nonassociative product problem mentioned above.

PROBLEM 2. Consider a random walk in the plane, where the steps are from (x,y) to $(x+1,y+1)$ or $(x+1,y-1)$, starting at a given point. In how many ways can the random walk go from $(0,0)$ to $(2n,0)$ through the upper halfplane without crossing the X-axis? Similarly we can demand that the walk does not meet the X-axis between $(0,0)$ and $(2n,0)$.

PROBLEM 3. A *tree* on n vertices is a connected graph with n vertices and $n-1$ edges. Such a graph is planar. If the graph is drawn in the plane we refer to it as a *plane tree*. A *rooted tree* is a tree with a distinguished vertex r called the *root*. If the valency of the root is 1 we say the tree is a *planted tree*. How many planted plane trees are there with n vertices?

PROBLEM 4. A planted plane tree is called *trivalent* (or *binary* tree or *bifurcating* tree) if every vertex has valency 1 or 3. It is easily seen that if there are n vertices of valency 1 then there are $n-2$ vertices of valency 3. How many trivalent planted plane trees are there with n vertices of valency 1 ?

PROBLEM 5. In how many ways can one decompose a convex $(n+1)$-gon into triangles by $n-2$ nonintersecting diagonals?

PROBLEM 6. In how many ways can 2n points on a circle be joined by n nonintersecting chords?

PROBLEM 7. A less familiar problem is the following. Let A_n be the set of n-tuples $(a_1,a_2,\dots,a_n)$ of integers > 1 such that in the sequence $1,a_1,a_2,\dots,a_n,1$ every a_i divides the sum of its two neighbors. Let U_n be defined in the same way, replacing > 1 by ≥ 1. Determine $|A_n|$ and $|U_n|$.

Of the very many references for these seven problems we list a few. Problem 1: [1], [2]; Problem 2: [3] Ch. 3; Problems 3 and 4: [4], [5]; Problem 5: [6].
Before showing the equivalence of the problems 1 to 7 we solve Problem 2 by a combinatorial argument. The method used is due to D. André and is called the *reflection principle* (cf. [3]). Let A and B be two points in the upper halfplane as in figure 9, and consider a path from A to B, which meets (or crosses) the X-axis.

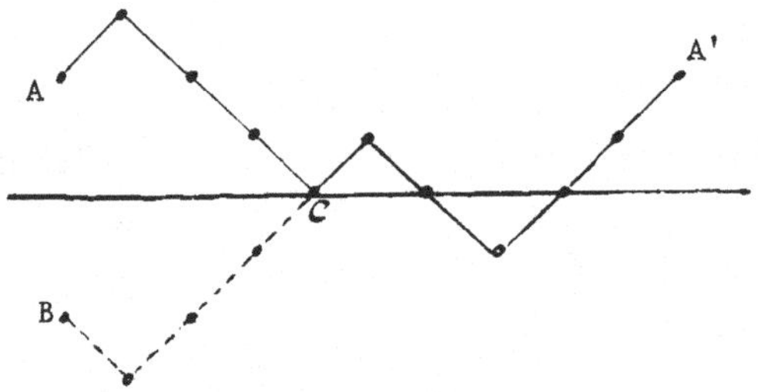

Fig. 9

By reflecting the part of the path between A and the first meeting with the X-axis (C in figure 9) with respect to the X-axis, we find a path from the reflected point A' to B. This establishes a 1-1 correspondence between paths from A' to B and paths from A to B which meet or cross the X-axis.
It follows that if A = (0,k) and B = (n,m), then there are $\binom{n}{\ell_1}$ paths from A to B which cross or meet the X-axis, where $2\ell_1 := n-k-m$. Since there are $\binom{n}{\ell_2}$ paths from A to B, where $2\ell_2 := n-m+k$, we find $\binom{n}{\ell_2} - \binom{n}{\ell_1}$ paths from A to B which do not meet the X-axis. Any path from (0,0) to (2n,0) through the upper halfplane which does not meet the X-axis between these points goes from (0,0) to (1,1) =: A, from A to B := (2n-1,1) without meeting the X-axis, and then from (2n-1,1) to (2n,0). By the argument above there are u_n such paths. If we allow the paths to meet the X-axis, without crossing, then there are u_{n+1} such paths. It seems very hard to find this number by a combinatorial argument which yields the factor n^{-1} (resp. $(n+1)^{-1}$) in a natural way.
We remark that the number of paths from (0,0) to (2n,0) through the upper halfplane which do not meet the X-axis between these points is equal to the number of sequences of zeros and ones

$$(x_1,x_2,\dots,x_{2n})$$

with

(3.1.2) $\qquad x_1 + x_2 + \ldots + x_j < \frac{1}{2}j$, $\qquad j = 1,2,\ldots,2n-1$,

(3.1.3) $\qquad x_1 + x_2 + \ldots + x_{2n} = n$.

The correspondence is given by letting a 1 correspond to a step $(x,y) \rightarrow (x+1,y-1)$ of the path.

We now turn to the problem of showing the equivalence of problems 1 to 7. In most cases we do not give a formal proof but simply illustrate the correspondence by a figure.

(i) *Problems 1 and 4*: The correspondence is shown by figure 10.

$$x_1((x_2x_3)(x_4x_5)) \;\rightleftarrows$$

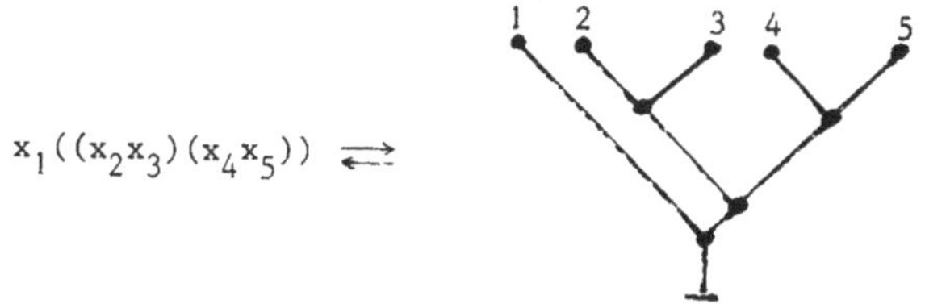

Fig. 10

It follows that the solution to Problem 4 is u_{n-1}.

(ii) *Problems 2 and 4*: Consider the trivalent planted plane tree in figure 11.

∙∙∙	level 4
∙∙∙	level 3
∙∙∙	level 2
∙∙∙	level 1
∙∙∙	level 0

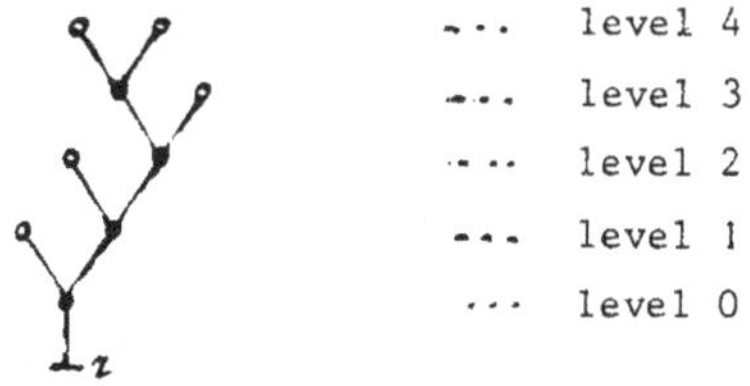

Fig. 11

We have ordered the vertices in *levels* and in each level we read from left to right. We describe the tree by a sequence of zeros and ones, taking a 0 for a vertex of valency 3, a 1 otherwise. We find 0 1 0 1 0 0 1 1 1. If we add a 0 in front (corresponding to the root) we have a sequence as in (3.1.2), (3.1.3) with n = 5. That (3.1.3) is satisfied is obvious and (3.1.2) follows from the fact that $(x_1,x_2,\ldots,x_j)$ is a sequence describing a partial tree corresponding to a lower part of figure 11. To finish the sequence a number of ones would have to be added. E.g. 0 0 1 0 1... and 0 0 1 0 1 1 correspond to figure 12:

Fig. 12

(iii) *Problems 2 and 3*: Consider a planted plane tree as in figure 13. Again the vertices are in levels and numbered in the obvious way.

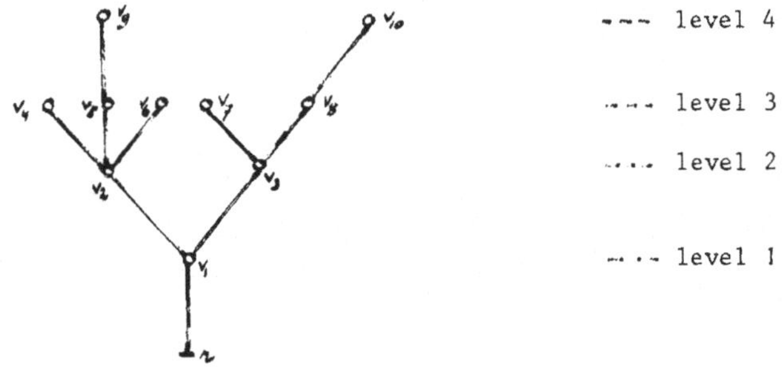

Fig. 13

We describe this tree by a sequence of e's (for edge) and the vertices v_i, each followed by as many e's as there are edges going up from v_i:

$$e\,v_1\,e\,e\,v_2\,e\,e\,e\,v_3\,e\,e\,v_4\,v_5\,e\,v_6\,v_7\,v_8\,e\,v_9\,v_{10}\ .$$

If we now replace each e by 0, each v_i by 1 we have a sequence $x_1, x_2, \ldots, x_{20}$ with $x_1 + x_2 + \ldots + x_{20} = 10$ and $x_1 + x_2 + \ldots + x_j \le \frac{1}{2}j$ for $j = 1, \ldots, 20$. This corresponds to the first question in Problem 2. Another mapping giving a correspondence with the second question in Problem 2 is given by the "up-down" code. This code is discussed in [4]. The idea is shown in figure 14.

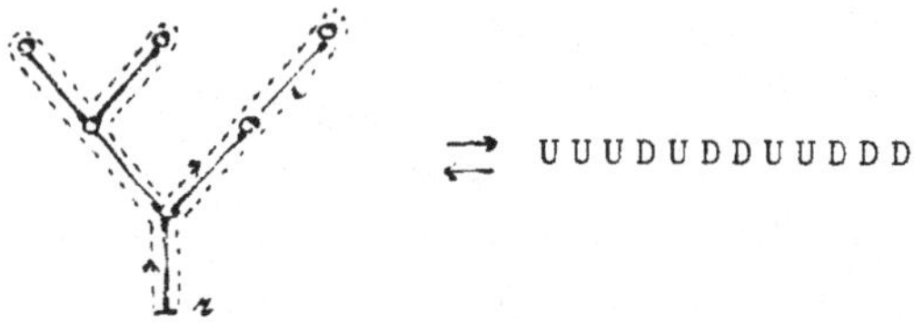

Fig. 14

The dotted line describes a path around the tree. For each edge, U (up) or D (down) gives the direction of the path. Clearly the number of U's exceeds the number of D's at every stage except when the path is complete. This corresponds to (3.1.2), (3.1.3) by taking U = 0, D = 1.

(iv) *Problems 3 and 4*: We take this correspondence from [4]. As the authors of [4] say: "The principle is so simple that it seems to be a pity to obscure it by giving a formal description ...". See figure 15.

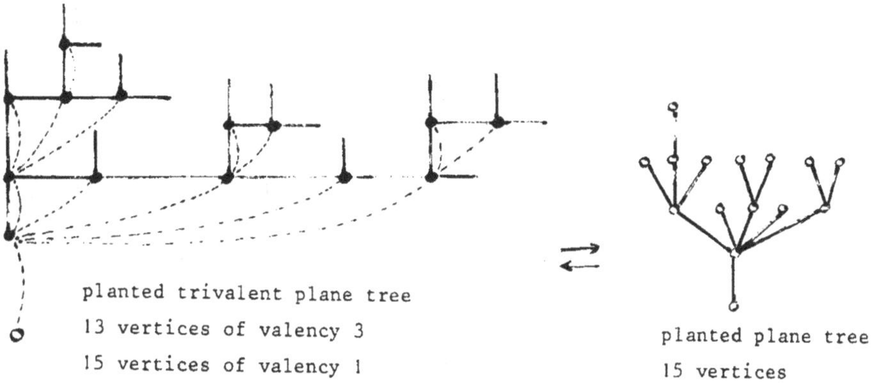

Fig. 15

(v) *Problems 1, 4 and 5*: We distinguish an edge of an (n+1)-gon, then consider the (n+1)-gon, decomposed into triangles as a planar graph and draw a modified "dual" graph of this graph. The rule is demonstrated in figure 16a. Subsequent application of the mapping discussed in (i) yields figure 16b.

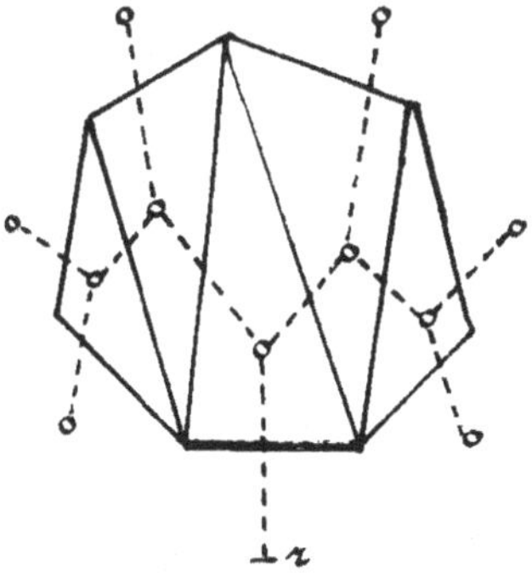

Fig. 16a

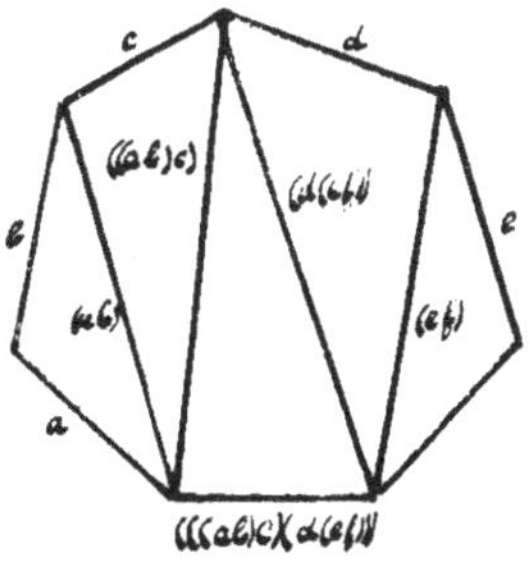

Fig. 16b

Clearly the "dual" graph in figure 16a is a trivalent plane tree which becomes planted if we consider the edge crossing the distinguished edge of the (n+1)-gon as coming from the root. (Note that the usual concept of dual graph is the same as ours if we identify all vertices of valency 1.) In this case a decomposition of an (n+1)-gon corresponds to a trivalent planted plane tree with n + 1 vertices of valency 1. Hence the solution to problem 5 is u_n.

(vi) *Problems 3 and 6*: Again a figure (figure 17) illustrates the equivalence. We leave a formal proof to those readers who are not convinced by figures.

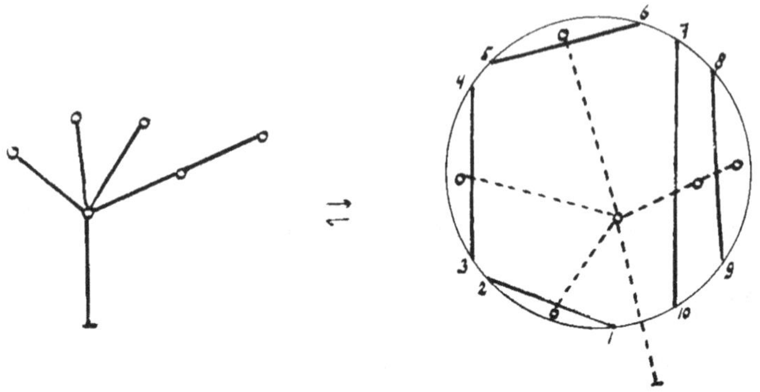

Fig. 17

The chords divide the circle into n parts. For each we have a vertex. The root is outside the circle and the edge from the root crosses the circle between 1 and 2n. The tree has n + 1 edges. Hence the solution to problem 6 is u_{n+1}.

Problems 7 and 2: To show the correspondence with the previous combinatorial problems we analyze a sequence $1, a_1, a_2, \ldots, a_n, 1$ as described in Problem 7. If, for any i, we have $a_i = a_{i+1}$ then the divisibility condition for the integers a_j implies that all of them are divisible by a_i and hence all the a_j's are 1. Otherwise there is at least one a_i such that $a_{i-1} < a_i$ and $a_i > a_{i+1}$. Then $a_i \mid (a_{i-1} + a_{i+1})$ implies that $a_{i-1} + a_{i+1} = a_i$ (we take $a_0 = a_{n+1} = 1$). It is easily checked that we can now remove a_i from the sequence, thus obtaining a sequence with one element less which still satisfies the divisibility condition. Conversely, any sequence can be lengthened by adding the term $a_i + a_{i+1}$ between a_i and a_{i+1}. For example

$$(1,1) \to (1,2,1) \to (1,2,3,1) \to (1,2,5,3,1) \to (1,2,5,3,4,1)$$

or

$$(1,1) \to (1,2,1) \to (1,2,3,1) \to (1,2,3,4,1) \to (1,2,5,3,4,1) .$$

We repeat this example, but now when a term $a_i + a_{i+1}$ is added between a_i and a_{i+1} we insert a mark before a_i and are allowed to make subsequent changes after the mark only. The second of the sequences does not satisfy this condition. The first example becomes

$$(1,1) \to (|1,2,1) \to (|1|2,3,1) \to (|1||2,5,3,1) \to (|1||2,5|3,4,1) .$$

The places of the 4 marks completely determine the sequence a_1, a_2, a_3, a_4 (in this case 2,5,3,4) and obviously the marks precede the corresponding numbers. The sequence starts with a mark. If we replace the sequence of marks and a_i's by 0's and 1's respectively and then omit the last two zeros, we have shown the correspondence with the first question in Problem 2. To show that this is indeed a 1-1 correspondence we note that a given sequence $a_1, a_2, \ldots, a_n$ can be reduced inductively by subsequently removing the term a_i with $a_{i-1} + a_{i+1} = a_i$, i maximal. This reverses the procedure described above.

The sequences forming the set U_n are treated in the same way, starting from (0,1,0). In this case we have three more integers than marks. Our rule says these are at the end. An example illustrates the procedure. Start from a sequence of 0's and 1's as in Problem 2, say 0 0 1 0 1 1. This corresponds to

$$|\,|0|\ ?\ ?\ ?\ 1\ 0\ ,$$

where we have added three more integers at the end, of which 1 and 0 are known. This describes the sequence generated as follows:

$$(0,1,0) \to (|0,1,1,0) \to (|\,|0,1,1,1,0) \to (|\,|0|1,2,1,1,0)\ .$$

We have shown that

$$(3.1.4) \qquad |U_n| = |A_{n+1}| = u_{n+2}\ .$$

3.2. *Stirling numbers.*

We recall the definitions of the Stirling numbers as given in H.Ch.3, Problem 2. We have

$$(3.2.1) \qquad (x)_0 := 1\ ,$$

$$(3.2.2) \qquad (x)_n := x(x-1)\ \ldots\ (x-n+1)\ , \qquad (n \in \mathbb{N})\ .$$

Then the Stirling numbers of the first kind $s(n,r)$ are defined by

$$(3.2.3) \qquad (x)_n =: \sum_{r=0}^{n} s(n,r)x^r \qquad (n \geq 0)\ .$$

The Stirling numbers of the second kind $S(n,r)$ are defined by

$$(3.2.4) \qquad x^n =: \sum_{r=0}^{n} S(n,r)(x)_r \qquad (n \geq 0)\ .$$

It is often useful to extend these definitions by defining $s(n,r) = S(n,r) = 0$ if $r < 0$ or $r > n$ (e.g. in H.p.27, Problem 3).

Generating functions. From (3.2.3) we find, for $|z| < 1$,

$$(1+z)^x = \sum_{n=0}^{\infty} \binom{x}{n} z^n = \sum_{n=0}^{\infty} \frac{1}{n!} (x)_n\, z^n =$$

$$= \sum_{n=0}^{\infty} \frac{1}{n!} z^n \sum_{r=0}^{n} s(n,r)x^r = \sum_{r=0}^{\infty} x^r \sum_{n=r}^{\infty} s(n,r) \frac{z^n}{n!}\ .$$

On the other hand, we have

$$(1 + z)^x = e^{x \log(1+z)} = \sum_{r=0}^{\infty} \frac{1}{r!} (\log(1 + z))^r x^r .$$

Hence it follows that

$$\sum_{n=r}^{\infty} s(n,r) \frac{z^n}{n!} = \frac{1}{r!} (\log(1 + z))^r . \tag{3.2.5}$$

In the same way we find from (3.2.4)

$$e^{xz} = \sum_{n=0}^{\infty} \frac{x^n z^n}{n!} = \sum_{n=0}^{\infty} \frac{z^n}{n!} \sum_{r=0}^{n} S(n,r)(x)_r = \sum_{r=0}^{\infty} (x)_r \sum_{n=r}^{\infty} S(n,r) \frac{z^n}{n!} .$$

Since we also have

$$e^{xz} = (1 + (e^z - 1))^x = \sum_{r=0}^{\infty} (x)_r \frac{(e^z - 1)^r}{r!} ,$$

we find that

$$\sum_{n=r}^{\infty} S(n,r) \frac{z^n}{n!} = \frac{1}{r!} (e^z - 1)^r . \tag{3.2.6}$$

For different proofs of (3.2.5) and (3.2.6) see [7].

Relations. The Stirling numbers of the first and second kind are connected by the relation

$$\sum_r S(n,r)s(r,m) = \delta_{nm} . \tag{3.2.7}$$

This immediately follows from (3.2.4) by substituting (3.2.3). Now we interpret this using the terminology of H. § 2.2. Let $P := \mathbb{N} \cup \{0\}$ with the usual ordering reversed. Then the functions s and S are elements of the incidence algebra A(P) of P. Since $S(n,n) = 1$ for $n \geq 0$, we find from H. Lemma 2.2.1 that S has an inverse $S^{\leftarrow}$, i.e.

$$\sum_{n \geq r \geq m} S(n,r)S^{\leftarrow}(r,m) = \delta_{nm} .$$

Apparently s is the inverse of S in A(P). If $(a_n)_{n \in \mathbb{N}}$ and $(b_n)_{n \in \mathbb{N}}$ are sequences, we define the functions a and b in A(P) by

$$a(x,y) := a_{x-y} \qquad (x \geq y) ,$$

$$b(x,y) := b_{x-y} \qquad (x \geq y) .$$

By (H.2.2.1) we can then interpret a relation

$$a_n = \sum_r s(n,r)b_r \qquad (n = 1,2,\ldots)$$

as

$$a = sb \; .$$

This implies

$$b = \overset{\leftarrow}{s} a = Sa \; ,$$

i.e.

$$b_n = \sum_r S(n,r) a_r \; .$$

This explains the relations (a), (b) of H.Ch.2 Problem 2 in terms of incidence algebras.

We now return to the formula (3.2.6). Expand the right-hand side and then expand e^{kz} in a power series and change the order of summation. This yields

$$r! \sum_{n=r}^{\infty} S(n,r) \frac{z^n}{n!} = \sum_{k=0}^{r} \binom{r}{k} (-1)^{r-k} e^{kz} =$$

$$= \sum_{k=0}^{r} \binom{r}{k} (-1)^{r-k} \sum_{n=0}^{\infty} k^n \frac{z^n}{n!} =$$

$$= \sum_{n=0}^{\infty} \frac{z^n}{n!} \sum_{k=0}^{r} (-1)^{r-k} \binom{r}{k} k^n \; .$$

It follows that

$$\text{(3.2.8)} \qquad \sum_{k=0}^{r} (-1)^{r-k} \binom{r}{k} k^n = \begin{cases} r! \; S(n,r) & (n \geq r) \; , \\ 0 & (n < r) \; . \end{cases}$$

The special case $r = n$ was treated in (2.1.6). We take a second look at (2.1.7). Apply (3.2.4):

$$f_n(x) = \sum_{k=0}^{\infty} k^n x^k = \sum_{k=0}^{\infty} x^k \sum_{r=0}^{n} S(n,r)(k)_r =$$

$$= \sum_{r=0}^{n} S(n,r) \sum_{k=0}^{\infty} (k)_r \, x^k = \sum_{r=0}^{n} S(n,r) r! \; x^r (1-x)^{-r-1} \; .$$

Hence $f_n(x) = (1-x)^{-n-1} \, P_n(x)$ where

$$\text{(3.2.9)} \qquad P_n(x) = \sum_{r=0}^{n} S(n,r) r! \; x^r (1-x)^{n-r}$$

and again we find $P_n(1) = n!$.

Combinatorial interpretations. There are a number of combinatorial interpretations of the Stirling numbers of the second kind. We shall consider these below. For the sake of completeness we remark that s(n,r) is the number of permutations of n symbols which have exactly r cycles (cf. [7] Ch.4.3).
Consider all the permutations of b_1 1's, b_2 2's, ..., b_r r's. Their number is the multinomial coefficient $\frac{n!}{b_1! \dots b_r!}$ (cf. (H.2.1.20)). Now, we wish to count the number of permutations (with repetition) of n symbols chosen from $x_1,x_2,\dots,x_r$ with the property that each symbol occurs at least once. Clearly this is the coefficient of $\frac{t^n}{n!}$ in the expansion of $(\frac{t}{1!} + \frac{t^2}{2!} + \dots)^r$. Hence, by (3.2.6) we have

THEOREM 3.2.1. *The number of permutations of* r *things taken* n *at a time, repeats permitted, such that each of the* r *things occurs at least once, is* r! S(n,r).

This can also be formulated as follows.

THEOREM 3.2.2. *The number of ways* n *distinct objects can be divided over* r *distinct boxes, with no box empty, is* r! S(n,r).

Proof. Let $o_1,o_2,\dots,o_n$ be the objects and number the boxes $x_1,x_2,\dots,x_r$. Consider one of the permutations $a_1,a_2,\dots,a_n$ counted by Theorem 3.2.1. This permutation corresponds to a division of the objects over the boxes in which o_i is in box x_j if $a_i = x_j$ $(i = 1,2,\dots,n)$. This is clearly a 1-1 correspondence.

COROLLARY. S(n,r) *is the number of ways of partitioning a set of* n *elements into* r *nonempty subsets.*

Proof. This follows from Theorem 3.2.2 by disregarding the order of the boxes.

Remark. If we also no longer consider the n elements of the set as distinguishable, then the number of partitions is $p_r(n)$ (cf. H.Ch.4). (For further results see [7] Ch.5.)

Two recent problems. We apply the results of this section to two interesting problems which appeared in Elemente der Mathematik.

PROBLEM 1 (El. d. Math. 27 (1972), Aufgabe 654, p.110). Show that

$$(3.2.10) \qquad S(n+r,n) = \sum_{1\le k_1\le \dots \le k_r \le n} k_1 k_2 \dots k_r \qquad (k_i \in \mathbb{N}) .$$

(This is a different formulation from the one which originally appeared.)

First solution. Apply the corollary of Theorem 3.2.2. Let the elements be $x_1,x_2,\dots,x_{n+r}$. If we divide $\{x_1,x_2,\dots,x_{n+r-1}\}$ into n subsets then there are n choices for the place of x_{n+r}. We can also let $\{x_{n+r}\}$ be one of the subsets and then divide $\{x_1,x_2,\ldots,x_{n+r-1}\}$ into $n-1$ subsets. It follows that

(3.2.11) $$S(n+r,n) = n\,S(n+r-1,n) + S(n+r-1,n-1)\ .$$

Let $F(r,n)$ be the right-hand side of (3.2.10). Divide the sum into two parts: (i) the terms with $k_r = n$ and (ii) the terms with $k_r \le n-1$. It follows that

(3.2.12) $$F(r,n) = n\,F(r-1,n) + F(r,n-1)\ .$$

Now, (3.2.10) follows by induction from (3.2.11) and (3.2.12) since the two sides are obviously equal for $r = 1$. (We could also define $F(0,n) := 1$, in which case (3.2.12) remains correct for $r = 1$.)

Second solution. Instead of using recursion we can also prove (3.2.10) by dividing the partitions counted by $S(n+r,n)$ into classes, each of which is counted by one term on the right-hand side of (3.2.10). Let the set $\{x_1,x_2,\ldots,x_{n+r}\}$ be partitioned into n nonempty subsets. We label a subset by the minimal i such that x_i is in the subset. Then order the labels: $1 = a_1 < a_2 < \ldots < a_n \le n+r$. Let $b_1 < b_2 < \ldots < b_r$ be the remaining x_i's. Define $k_i := |\{j \in \mathbb{N} \mid a_j < b_i\}| = b_i - i$. Then the numbers $b_1,b_2,\ldots,b_r$ can be divided over the subsets in $k_1\,k_2\,\ldots\,k_r$ ways in accordance with the labeling. It is easily seen that the sequence $k_1,k_2,\ldots,k_r$ satisfies $1 \le k_1 \le k_2 \le \ldots \le k_r \le n$ and that any such a sequence uniquely determines a sequence of labels $1 = a_1 < a_2 < \ldots < a_n \le n+r$. This proves (3.2.10).

PROBLEM 2 (El. d. Math. 27, 1972, Aufgabe 673, p.95). Let Φ denote a permutation of $1,2,\ldots,n$ and let $F(\Phi)$ denote the number of fixed points of Φ. Show that

(3.2.13) $$A(n,k) := \frac{1}{n!}\sum_{\Phi} (F(\Phi))^k = A_k\ ,$$

where A_k is the number of partitions of $\{1,2,\ldots,k\}$ and the summation is over all permutations of $\{1,2,\ldots,n\}$.

Solution. Let D_m be the number of derangements of m symbols. Then

$$A(n,k) = \frac{1}{n!}\sum_{m=0}^{n} \binom{n}{m} D_{n-m}\, m^k$$

and by (2.2.5) this is the coefficient of x^n in the product

(3.2.14) $$e^{-x}(1-x)^{-1}\sum_{m=0}^{\infty} m^k \frac{x^m}{m!}\ .$$

An immediate consequence of (3.2.11) is

(3.2.15) $$e^x \sum_{\ell=1}^{k} S(k,\ell)x^\ell = \left(x\frac{d}{dx}\right)^k e^x = \sum_{m=0}^{\infty} m^k \frac{x^m}{m!}\ .$$

(3.2.14) and (3.2.15) imply that $A(n,k)$ is the coefficient of x^n in

$(1-x)^{-1} \sum_{\ell=1}^{k} S(k,\ell)x^{\ell}$ and for $n \geq k$ this coefficient is $\sum_{\ell=1}^{k} S(k,\ell)$. By the corollary to Theorem 3.2.2 this is equal to A_k.

The material of this section gives an impression of the many connections between the Stirling numbers and several of the topics treated in H.Ch.1,2,3. For much more on Stirling numbers we refer to [7].

References.

[1] H.W. Becker, Discussion of Problem 4277, Am. Math. Monthly 56 (1949), 697-699.

[2] O. Ore, Problem 3954, Am. Math. Monthly 48 (1941), 564-569.

[3] W. Feller, An Introduction to Probability Theory and its Applications, John Wiley & Sons, New York, 1950.

[4] N.G. de Bruijn and B.J.M. Morselt, A Note on Plane Trees, J. of Comb. Theory 2 (1967), 27-34.

[5] D.A. Klarner, Correspondences between Plane Trees and Binary Sequences, J. of Comb. Theory 9 (1970), 401-411.

[6] E. Lucas, Théorie des Nombres, Paris, 1891.

[7] J. Riordan, An Introduction to Combinatorial Analysis, John Wiley & Sons, New York, 1958.

[8] H.W. Gould, Research Bibliography of Two Special Number Sequences, Mathematicae Monongaliae (Dept. of Math., W.Va. Univ. 26506), 1971.

IV. PARTITIONS

The sessions devoted to chapter IV did not yield many results. In this chapter we shall make some comments on $p_3(n)$ (cf. (H.4.1.6)), on asymptotic properties of $p(n)$ (cf. (H.4.2.8)), and a few remarks concerning special partition problems (H.p.43).

4.1. *The number* $p_3(n)$.

Let $a_3(n)$ be the number of solutions of

(4.1.1) $$n = x_1 + x_2 + x_3 \,, \qquad x_1 \ge x_2 \ge x_3 \ge 0 \,.$$

Then $a_3(n) = p_3(n+3)$ (cf. (H.4.1.4)). Now write $y_3 := x_3$, $y_2 := x_2 - x_3$, $y_1 := x_1 - x_2$. Then we find that $a_3(n)$ is the number of solutions of

(4.1.2) $$n = y_1 + 2y_2 + 3y_3 \,, \qquad y_i \ge 0,\ i = 1,2,3 \,.$$

It follows that

(4.1.3) $$\sum_{n=0}^{\infty} a_3(n)x^n = (1-x)^{-1}(1-x^2)^{-1}(1-x^3)^{-1} \,.$$

By partial fraction decomposition of the right-hand side of (4.1.3) we find

(4.1.4) $$\sum_{n=0}^{\infty} a_3(n)x^n = \frac{1}{6}(1-x)^{-3} + \frac{1}{4}(1-x)^{-2} + \frac{17}{72}(1-x)^{-1} + \frac{1}{8}(1+x)^{-1} + \frac{1}{9}(1-\omega x)^{-1} + \frac{1}{9}(1-\omega^2 x)^{-1} \,,$$

where ω and ω^2 are the two nonreal cube roots of unity. Each of the terms on the right-hand side of (4.1.4) can be expanded in a power series. Combining the series we find

(4.1.5) $$a_3(n) = \frac{1}{12}(n+3)^2 - \frac{7}{72} + \frac{(-1)^n}{8} + \frac{1}{9}(\omega^n + \omega^{2n}) \,.$$

Therefore

$$\left|a_3(n) - \frac{1}{12}(n+3)^2\right| \le \frac{7}{72} + \frac{1}{8} + \frac{2}{9} < \frac{1}{2} \,,$$

which implies

(4.1.6) $$p_3(n) = \left\langle \frac{1}{12} n^2 \right\rangle \,,$$

(the integer nearest to $\frac{1}{12}n^2$).

The six formulas for $p_3(n)$ given in (H.4.1.6) are in accordance with this result. This derivation of $p_3(n)$ occurs a.o. in [1] and also in [2] as the solution of the following problem of N. Anning.

PROBLEM. From the vertices of a regular n-gon three are chosen to be the vertices of a triangle. Prove that the number of essentially different possible triangles is the integer nearest to $\frac{n^2}{12}$.

It is easily seen that the required number is $p_3(n)$ and then the solution follows from (4.1.6). It is also possible to count the triangles directly, thus giving an alternative proof of (4.1.6) (cf. [2]).
Of course the method used to derive (4.1.6) can also be used for $p_k(n)$, $k > 3$, but the results are much more complicated, e.g.

(4.1.7) $$\sum_{n=0}^{\infty} a_4(n)x^n = \frac{1}{24}(1-x)^{-4} + \frac{1}{8}(1-x)^{-3} + \frac{59}{288}(1-x)^{-2} + $$
$$+ \frac{17}{72}(1-x)^{-1} + \frac{1}{32}(1+x)^{-2} + \frac{1}{8}(1+x)^{-1} + \frac{1-\omega}{27}(1-\omega x)^{-1} + $$
$$+ \frac{1-\omega^2}{27}(1-\omega^2 x)^{-1} + \frac{1}{16}(1-ix)^{-1} + \frac{1}{16}(1+ix)^{-1},$$

(4.1.8) $$a_4(n) = \frac{1}{144}(n+1)(n+2)(n+3) + \frac{1}{16}(n+1)(n+2) + \frac{59}{288}(n+1) + $$
$$+ \frac{17}{72} + \frac{(-1)^n(n+1)}{32} + \frac{1}{8}(-1)^n + \frac{1-\omega}{27}\omega^n + \frac{1-\omega^2}{27}\omega^{2n} + \frac{1}{16}i^n + \frac{1}{16}(-i)^n,$$

or

(4.1.9) $$a_4(n) = \left\lceil \frac{1}{24}\binom{n+3}{3} + \frac{1}{8}\binom{n+2}{2} + \frac{59+9(-1)^n}{288}(n+1) \right\rceil .$$

(This solves Problem 6 on H.p.43.) (Here $\lceil x \rceil := \min\{n \in \mathbb{N} \mid n \geq x\}$.)

4.2. *Asymptotic properties of* p(n).

For p(n), the number of unrestricted partitions of n, M. Hall mentions the result

(4.2.1) $$p(n) \sim \frac{1}{4n\sqrt{3}} e^{\pi\sqrt{\frac{2}{3}n}} \qquad (n \to \infty),$$

(cf. (H.4.2.8)) but no proof is given in the book since the proof is long and more typically a part of analytic number theory than of combinatorics.
It seems worthwhile to give a short proof of an inequality which strongly resembles (4.2.1). The proof is taken from [1].

THEOREM 4.2.1.

(4.2.2) $$p(n) < e^{\pi\sqrt{\frac{2}{3}n}} .$$

Proof. Let

(4.2.3) $$f(t) := \sum_{n=0}^{\infty} p(n)t^n = \prod_{k=1}^{\infty} (1 - t^k)^{-1} \qquad (|t| < 1) .$$

For $0 < t < 1$ we define $g(t) := \log f(t)$. Then we have

(4.2.4) $$g(t) = - \sum_{k=1}^{\infty} \log(1 - t^k) = \sum_{k=1}^{\infty} \sum_{j=1}^{\infty} \frac{t^{kj}}{j} = \sum_{j=1}^{\infty} \frac{t^j}{1 - t^j} .$$

From $(1 - t)^{-1}(1 - t^j) = 1 + t + \ldots + t^{j-1}$ it follows that

$$jt^{j-1} < (1-t)^{-1}(1-t^j) < j \qquad (0 < t < 1,\ j \ge 1) .$$

Therefore we have from (4.2.4)

$$\sum_{j=1}^{\infty} \frac{t^{j-1}}{j^2} < \frac{(1-t)}{t} g(t) < \sum_{j=1}^{\infty} j^{-2} = \frac{1}{6}\pi^2$$

and hence

(4.2.5) $$\lim_{t \uparrow 1} \frac{(1-t)}{t} g(t) = \frac{1}{6}\pi^2 .$$

The definition of $f(t)$ implies that $f(t) > p(n)t^n$ for $0 < t < 1$, $n \in \mathbb{N}$. Hence

(4.2.6) $$\log p(n) + n \log t < g(t) < \frac{1}{6}\pi^2 \frac{t}{1-t} .$$

We substitute $t = (1+u)^{-1}$ $(0 < u < \infty)$ and find

$$\log p(n) < \frac{1}{6}\pi^2 u^{-1} + n \log(1+u) < \frac{1}{6}\pi^2 u^{-1} + nu ,$$

in which the final expression has its minimal value for $u = \pi(6n)^{-\frac{1}{2}}$. Substituting this value of u we find the required result.

Of course this inequality is much weaker than (4.2.1). However, we now show that a slight modification of the proof gets us much closer to (4.2.1). Clearly $p(n)$ increases with n. Therefore at the point where (4.2.6) is derived we can also conclude from (4.2.3) that $f(t) > p(n)t^n(1-t)^{-1}$. Again we take logarithms and substitute $t = (1+u)^{-1}$. The result is

$$\log p(n) < \frac{1}{6}\pi^2 u^{-1} + n \log(1+u) + \log \frac{u}{1+u}$$
$$< \frac{1}{6}\pi^2 u^{-1} + (n-1)u + \log u .$$

Now we substitute $u = \dfrac{-1 + (1 + \frac{2}{3}\pi^2(n-1))^{\frac{1}{2}}}{2(n-1)}$. Dropping insignificant terms it follows that

$$p(n) < \frac{\pi}{\sqrt{6(n-1)}} e^{\pi\sqrt{\frac{2}{3}n}} . \tag{4.2.7}$$

We remark that this improvement of (4.2.2) takes no more work than the proof of (4.2.2) itself!

4.3. *Partitions, series and products.*

It is shown in H.p.43 Problem 2 that p(n) is convex. We give a proof of this statement using series and products (cf. [3]). Let $c(0) := 1$ and $c(n) :=$ the number of partitions of n into parts ≥ 3. Then $c(1) = c(2) = 0$, $c(3) = 1$ and $c(n) \geq c(n-1)$ for $n \geq 2$. It follows that for $n \geq 2$ the coefficient of x^n in the expansion of

$$(1+x)^{-1} \sum_{n=0}^{\infty} c(n)x^n$$

is positive. Now

$$(1+x)^{-1} \sum_{n=0}^{\infty} c(n)x^n = (1-x)^2(1-x)^{-1}(1-x^2)^{-1} \prod_{k=3}^{\infty} (1-x^k)^{-1} =$$

$$= (1-x)^2 \prod_{k=1}^{\infty} (1-x^k)^{-1} = (1-x)^2 \sum_{n=0}^{\infty} p(n)x^n =$$

$$= p(0) + \{p(1) - 2p(0)\}x + \sum_{n=2}^{\infty} \{p(n) - 2p(n-1) + p(n-2)\}x^n$$

and therefore $p(n) - 2p(n-1) + p(n-2) \geq 0$ for $n \geq 2$.

Problem 4 on H.p.43 deserves a comment in the opposite direction, namely a combinatorial proof instead of one by analysis. The equality

$$\prod_{k=1}^{\infty} (1+x^k) = \prod_{\ell=1}^{\infty} (1-x^{2\ell-1})^{-1} \tag{4.3.1}$$

can be proved by showing that the number of partitions of n into distinct parts is equal to the number of partitions of n into odd parts. (Of course a direct proof of (4.3.1) is practically trivial.) This partition identity is demonstrated in a nice combinatorial way by mapping the partition $\ell_1 + \ell_2 \cdot 3 + \ell_3 \cdot 5 + \ldots$ with $\ell_i = \sum_{j=0}^{\infty} \varepsilon_{ij} 2^j$ into the partition $\sum_{i=1}^{\cdot} \sum_{j=0}^{\cdot} (2i-1)\varepsilon_{ij} 2^j$, since clearly no two nonzero terms in this finite double summation are equal. The mapping is obviously one to one.

The array of dots often used to describe a partition (e.g. in H. Theorem 4.1.3) is generally called a Ferrers graph. A pictorial demonstration of this type that the number of partitions of n into odd parts is equal to the number of partitions into unequal parts is given in figure 18a. The odd number of dots corresponding to a term in the partition is placed symmetrically with respect to the diagonal of the figure.

Fig. 18a

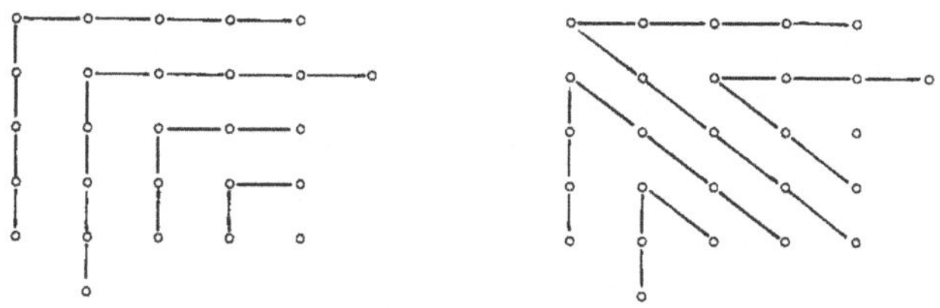

Fig. 18b

27 = 9 + 9 + 5 + 3 + 1 27 = 9 + 7 + 6 + 4 + 1

If we read the figure as is shown in figure 18b we find a partition into unequal parts. The reader can convince himself that this is a 1-1 mapping. The correspondence is due to Sylvester (cf. [4]).

To show that the correspondence is 1-1 consider the i-th component in figure 18b. Let α_i be the number of points in the horizontal (or vertical) part of this component and β_i the remaining number of points. (In figure 18b we have $\alpha_1 = 5$, $\beta_1 = 4$, $\alpha_2 = 4$, $\beta_2 = 3$, etc.) It is easily seen that $\alpha_{2i+1} = \alpha_{2i} - 1$, $\beta_{2i} = \beta_{2i-1} - 1$, and furthermore the sequences $\alpha_1, \alpha_2, \ldots$ and $\beta_1, \beta_2, \ldots$ are nonincreasing. In the final component $\alpha_i = 1$ if i is odd, resp. $\beta_i = 0$ if i is even. Therefore $\alpha_i + \beta_i$ is decreasing and furthermore the sequences $\alpha_1, \alpha_2, \ldots$ and $\beta_1, \beta_2, \ldots$ are determined by the sequence $\alpha_1 + \beta_1, \alpha_2 + \beta_2, \ldots$. They thus yield a configuration of the type of figure 18. This can also be expressed as follows. If $2a_1 + 1 > 2a_2 + 1 > \ldots$ are the components in figure 18a and if n_i denotes the number of integers $a_j \geq i$, then the partition of figure 18b is

$$n = (a_1 + n_0) + (a_1 - 1 + n_1) + (a_2 - 2 + n_1) + (a_2 - 3 + n_2) + (a_3 - 4 + n_2) + \ldots .$$

If we read a Ferrers graph (in the usual form) horizontally resp. vertically (cf. figure 19a) we find two partitions of n which are called *conjugate* partitions. If these are the same the partition is called selfconjugate (figure 19b).

Fig. 19a

8 = 4 + 3 + 1 = 3 + 2 + 2 + 1

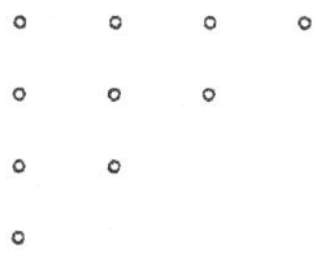

Fig. 19b

10 = 4 + 3 + 2 + 1

If we now read figure 19b in the same way as figure 18a we find 10 = 7 + 3, a partition into unequal odd parts. Again this is a 1-1 correspondence, this time between partitions of n into unequal odd parts and selfconjugate partitions of n. Now consider the partition $n = \ell_1 + \ell_2 + \ldots + \ell_s$, $\ell_1 < \ell_2 < \ldots < \ell_s$, where all the ℓ_i are odd and symbolize this by the usual Ferrers graph as in figure 20.

Fig. 20 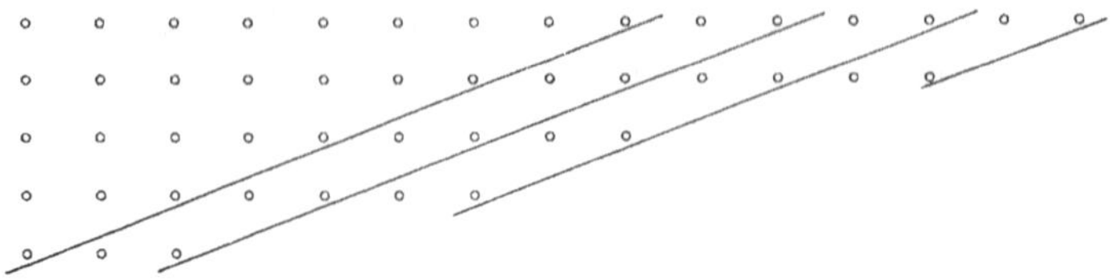

$47 = 3 + 7 + 9 + 13 + 15$, $47 = 25 + 10 + 8 + 4$

The lines in figure 20 correspond to the partition

$$n = [1 + 3 + 5 + \dots + (2s - 1)] + \frac{\ell_1 - 1}{2}\, 2s + \left(\frac{\ell_2 - \ell_1 - 2}{2}\right)(2s - 2) + \left(\frac{\ell_3 - \ell_2 - 2}{2}\right)(2s - 4) + \dots .$$

If we express this relation between partitions in series form we find

$$(4.3.2) \qquad \prod_{k=1}^{\infty} (1 + x^{2k-1}) = \sum_{s=0}^{\infty} x^{s^2} \prod_{\ell=1}^{s} (1 - x^{2\ell})^{-1} ,$$

which is H.p.43 Problem 4.

We give one more theorem of this type with a similar pictorial proof due to D.H. Lehmer (cf. [4]).

Consider a partition of n into k distinct parts (cf. figure 21).

Fig. 21 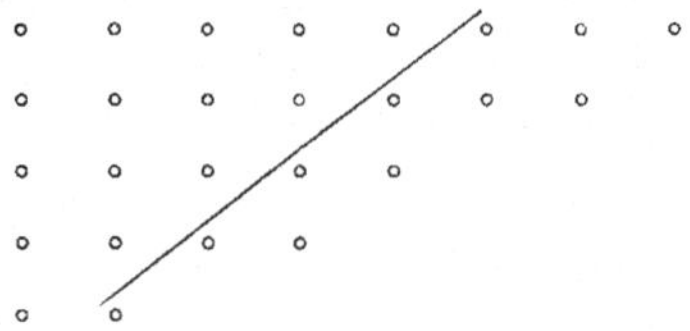

$26 = 8 + 7 + 5 + 4 + 2$

If we read the part of the figure to the right of the line by diagonals parallel to the line (in the figure $11 = 5 + 4 + 2$) we find a partition of $n - \frac{1}{2}k(k+1)$ into parts $\le k$. Again this is a 1-1 correspondence. Expressing this in product-series form we find

$$(4.3.3) \qquad \prod_{\ell=1}^{\infty} (1 + x^{\ell}) = \sum_{k=0}^{\infty} x^{\frac{1}{2}k(k+1)} \prod_{j=1}^{k} (1 - x^{j})^{-1} .$$

As a further reference we mention [5] Chapter 19.

References.

[1] K. Chandrasekharan, Arithmetical Functions, Springer Verlag, New York, 1970.

[2] N. Anning, Problem 3893, American Math. Monthly 47 (1940), 664-666.

[3] D.A. Klarner and J.H. van Lint, Problem 213, Nieuw Archief voor Wiskunde 18 (1970), 92.

[4] H.S. Wall, Problem 4067, American Math. Monthly 51 (1944), 353-355.

[5] G.H. Hardy and E.M. Wright, An Introduction to the Theory of Numbers, Oxford University Press, Oxford, 1945.

V. DISTINCT REPRESENTATIVES

The discussion of H.Ch.5 led to so many interesting questions that this report on our activities concerning that chapter will be quite long. The first question concerned the corollary to H. Theorem 5.1.1 which gives a lower bound for the number of SDR's of n sets $S_1, S_2, \dots, S_n$ (given that there is at least one SDR). This result which is due to M. Hall [1] was improved by Rado [2] in 1967. The question was raised whether further improvement was possible. In section 5.1 we present a proof of the best possible theorem in this direction which was found by M.L.J. Hautus and the author. Some time later our attention was drawn to a paper by P.A. Ostrand [3] which contained the same result with a proof which is nearly the same as ours.

5.1. *On the number of systems of distinct representatives of sets.*

In this section we consider finite sets. The number of elements of a set S is denoted by $|S|$.

DEFINITION 1. *A sequence* $(A_0, \dots, A_{n-1})$ *of subsets of a set* S *is said to have property H (P. Hall's condition) if for each* $k \in \{1,2,\dots,n\}$ *the union of any* k*-tuple of the* A_i*'s contains at least* k *elements.* (Condition C of H. § 5.1.)

DEFINITION 2. *If a sequence* $(A_0, \dots, A_{n-1})$ *of subsets of a set* S *has property H and if for some* $k \in \{1,2,\dots,n-1\}$ *there is a* k*-tuple of the* A_i*'s the union of which contains exactly* k *elements then we say the* k*-tuple is a critical block.*

DEFINITION 3. *Let* $(A_0, \dots, A_{n-1})$ *be a sequence of subsets of a set* S. *We shall say that* $(a_0, \dots, a_{n-1})$ *is a system of distinct representatives (SDR) of* $(A_0, \dots, A_{n-1})$ *if* $a_i \in A_i$ $(i = 0, \dots, n-1)$ *and* $a_i \neq a_j$ *if* $i \neq j$. *We denote by* $N(A_0, \dots, A_{n-1})$ *the number of SDR's of* $(A_0, \dots, A_{n-1})$.

DEFINITION 4. *Let* $m_0 \le m_1 \le \dots \le m_{n-1}$ *be a nondecreasing sequence of positive integers. We define:*

$$H_n(m_0, m_1, \dots, m_{n-1}) := \prod_{0 \le i < \min(m_0, n)} (m_0 - i) ,$$

$$G_n(m_0, m_1, \dots, m_{n-1}) := \prod_{0 \le i < \min(m_0, n)} (m_i - i) ,$$

$$F_n(m_0, m_1, \dots, m_{n-1}) := \prod_{0 \le i < n} (m_i - i)_* ,$$

where $(a)_* := \max\{1, a\}$.

Clearly

$$H_n(m_0, \dots, m_{n-1}) \le G_n(m_0, \dots, m_{n-1}) \le F_n(m_0, \dots, m_{n-1}) .$$

We shall now state what is known about the number of SDR's of a sequence of sets (not including the result of [3]).

THEOREM 5.1.1. *Let* $(A_0,\dots,A_{n-1})$ *be a sequence of subsets of a set* S. *Let* $m_i := |A_i|$ $(i = 0,1,\dots,n-1)$ *and let* $m_0 \le m_1 \le \dots \le m_{n-1}$. *If the sequence has property H, then*

(a) $N(A_0,\dots,A_{n-1}) \ge 1$, (P. Hall [4]),

(b) $N(A_0,\dots,A_{n-1}) \ge H_n(m_0,\dots,m_{n-1})$, (M. Hall [1]),

(c) $N(A_0,\dots,A_{n-1}) \ge G_n(m_0,\dots,m_{n-1})$, (R. Rado [2]).

Our refinement is

THEOREM 5.1.2. *Let* $(A_0,\dots,A_{n-1})$ *be a sequence of subsets of a set* S. *Let* $m_i := |A_i|$ $(i = 0,1,\dots,n-1)$ *and let* $m_0 \le m_1 \le \dots \le m_{n-1}$. *If the sequence has property H, then*

$$N(A_0,\dots,A_{n-1}) \ge F_n(m_0,\dots,m_{n-1}) .$$

Before proving Theorem 5.1.2 we show that this is the best lower bound for $N(A_0,\dots,A_{n-1})$ involving only $(m_0,\dots,m_{n-1})$ by showing that for every nondecreasing sequence $m_0 \le m_1 \le \dots \le m_{n-1}$ there is a sequence $(A_0,\dots,A_{n-1})$ of subsets of $\mathbb{N}$ with $|A_i| = m_i$ $(i = 0,1,\dots,n-1)$ which has exactly $F_n(m_0,\dots,m_{n-1})$ different SDR's. For this purpose we define:

$$A_i := \begin{cases} \{1,2,\dots,m_i\} & \text{if } m_i - i \ge 1 , \\ \{1,2,\dots,m_i-1,i+1\} & \text{if } m_i - i < 1 . \end{cases}$$

We choose the distinct reprresentatives successively. Clearly in A_0 we have m_0 choices. Since $A_0 \cup A_1 \cup \dots \cup A_{i-1} = \{1,2,\dots,\max\{m_{i-1},i\}\}$ we have $m_i - i$ choices in A_i if $m_i - i \ge 1$ and exactly one choice (namely $i+1$) if $m_i - i < 1$. Therefore there are $\prod_{i=0}^{n-1} (m_i - i)_* = F_n(m_0,\dots,m_{n-1})$ different SDR's of $(A_0,\dots,A_{n-1})$.

For the proof of Theorem 5.1.2 we need the following lemma:

LEMMA 5.1.1. *For* $n \ge 1$ *let* $f_n\colon \mathbb{Z}^n \to \mathbb{N}$ *be defined by*

$$f_n(a_0,a_1,\dots,a_{n-1}) := F_n(m_0,m_1,\dots,m_{n-1})$$

if $(m_0,m_1,\dots,m_{n-1})$ *is a nondecreasing rearrangement of* $(a_0,a_1,\dots,a_{n-1})$. *Then* f_n *is nondecreasing with respect to each of the variables* a_i.

Proof. Let $m_0 \le m_1 \le \dots \le m_{k-1} \le a_i \le m_{k+1} \le \dots \le m_\ell \le m_{\ell+1} \le \dots \le m_{n-1}$ be a nondecreasing rearrangement of $(a_0,a_1,\dots,a_{n-1})$. If $a_i' \ge a_i$ and $m_0 \le m_1 \le \dots \le m_{k-1} \le$ $\le m_{k+1} \le \dots \le m_\ell \le a_i' \le m_{\ell+1} \le \dots \le m_{n-1}$ is a nondecreasing rearrangement of

$(a_0,\dots,a_{i-1},a_i',a_{i+1},\dots,a_{n-1})$ then

$$\frac{f_n(a_0,\dots,a_{i-1},a_i',a_{i+1},\dots,a_{n-1})}{f_n(a_0,\dots,a_{n-1})} = \frac{(m_{k+1}-k)_*}{(a_i-k)_*} \prod_{j=k+1}^{\ell-1} \frac{(m_{j+1}-j)_*}{(m_j-j)_*} \frac{(a_i'-\ell)_*}{(m_\ell-\ell)_*}$$

and this is ≥ 1 since $a_i \leq m_{k+1}$, $a_i' \geq m_\ell$ and $m_{j+1} \geq m_j$ for $j = k+1,\dots,\ell-1$.

Proof of Theorem 5.1.2. Just as in H. and [2] we follow the idea of Halmos and Vaughan [5]. Clearly Theorem 5.1.2 is true for $n = 1$. Now let $n \geq 2$ and use induction with respect to n. We distinguish two cases:

Case 1. There is no critical block. Then (cf. H.) we may choose any element a of A_0 as its representative and delete this element from all other sets. This gives us the sets $A_1(a),\dots,A_{n-1}(a)$ for which condition *H* still holds and by the induction hypothesis Theorem 5.1.2 is true for this sequence of sets. We find (applying Lemma 5.1.1)

$$\begin{aligned} N(A_0,\dots,A_{n-1}) &\geq \sum_{a\in A_0} f_{n-1}(|A_1(a)|,\dots,|A_{n-1}(a)|) \geq \\ &\geq \sum_{a\in A_0} f_{n-1}(m_1-1,\dots,m_{n-1}-1) = \\ &= m_0\, f_{n-1}(m_1-1,\dots,m_{n-1}-1) = F_n(m_0,m_1,\dots,m_{n-1}) \;. \end{aligned}$$

Case 2. There is a critical block $(A_{\nu_0},A_{\nu_1},\dots,A_{\nu_{k-1}})$ with $\nu_0 < \nu_1 < \dots < \nu_{k-1}$ and $0 < k < n$. In this case (cf. H.) we delete all elements of $A_{\nu_0} \cup A_{\nu_1} \cup \dots \cup A_{\nu_{k-1}}$ from all the other sets A_i which produces $A'_{\mu_0},A'_{\mu_1},\dots,A'_{\mu_{\ell-1}}$ where $\{\nu_0,\nu_1,\dots,\nu_{k-1},\mu_0,\mu_1,\dots,\mu_{\ell-1}\} = \{0,1,\dots,n-1\}$, $k+\ell = n$.
Now both $(A_{\nu_0},A_{\nu_1},\dots,A_{\nu_{k-1}})$ and $(A'_{\mu_0},\dots,A'_{\mu_{\ell-1}})$ satisfy condition *H* and SDR's of the two sequences are always disjoint. Hence by the induction hypothesis and Lemma 5.1.1

$$\begin{aligned} (5.1.1) \qquad N(A_0,\dots,A_{n-1}) &= N(A_{\nu_0},\dots,A_{\nu_{k-1}})N(A'_{\mu_0},\dots,A'_{\mu_{\ell-1}}) \geq \\ &\geq f_k(m_{\nu_0},m_{\nu_1},\dots,m_{\nu_{k-1}})f_\ell(|A'_{\mu_0}|,\dots,|A'_{\mu_{\ell-1}}|) \geq \\ &\geq f_k(m_{\nu_0},m_{\nu_1},\dots,m_{\nu_{k-1}})f_\ell(m_{\mu_0}-k,m_{\mu_1}-k,\dots,m_{\mu_{\ell-1}}-k) \geq \\ &\geq f_k(m_0,m_1,\dots,m_{k-1})f_\ell(m_{\mu_0}-k,m_{\mu_1}-k,\dots,m_{\mu_{\ell-1}}-k) \;. \end{aligned}$$

Now we remark that

$$m_{\nu_{k-1}} \leq |A_{\nu_0} \cup \dots \cup A_{\nu_{k-1}}| = k \;,$$

and therefore we have

$$(m_r - r)_* = 1 \qquad \text{if } k \le r \le \nu_{k-1} ,$$

and

$$(m_{\mu_i} - k - i)_* = 1 \quad \text{if } \mu_i \le \nu_{k-1} .$$

This implies

$$f_k(m_0, m_1, \ldots, m_{k-1}) = \prod_{0 \le i \le \nu_{k-1}} (m_i - i)_* ,$$

$$f_\ell(m_{\mu_0} - k, \ldots, m_{\mu_{\ell-1}} - k) = \prod_{\nu_{k-1} < j < n} (m_j - j)_* ,$$

i.e. the product (5.1.1) is equal to $F_n(m_0, m_1, \ldots, m_{n-1})$ which proves the theorem.

Remark. Theorem 5.1.2 can be interpreted as an improvement of known lower bounds for permanents of (0,1)-matrices (cf. Ryser [6]). The formulation is as follows.

Corollary. Let A be a (0,1)-matrix of size m by n ($m \le n$) and let $r_1 \le r_2 \le \ldots \le r_m$ be the row sums in increasing order. Then if per(A) $\ne 0$ we have

$$\text{per}(A) \ge \prod_{i=1}^{m} (r_i - i + 1)_* .$$

5.2. *Independent representatives.*

In this section we present the contents of a research paper entitled "An algorithm proof of Rado's theorem on independent representatives" written by M.L.J. Hautus. The paper was presented in the seminar in connection with a question concerning SDR's for sets of vectors where linear independence of the SDR was required. The problem was raised by N.G. de Bruijn who also made several valuable suggestions. This section is made up of 3 parts, namely an introduction to the problem, a proof of the theorem and the algorithm.

Introduction

Rado [7] gave a generalization of P. Hall's [4] theorem on distinct representatives. The formulation of the theorem uses the concept of an independence relation. With a notation which is slightly different from the one used in [7] this concept is defined here.

DEFINITION 1. *Let* S *be a set. An independence relation on* S *is a sequence of relations* $I_1, I_2, \ldots$, *with* $I_n \subset S^n$ *(*S^n *is the* n*-th cartesian power of* S*, hence,* I_n *is an* n*-ary relation on* S*), such that the following properties hold:*

(5.2.1) $\qquad (x_1, \ldots, x_m) \in I_m \Rightarrow (x_1, \ldots, x_{m-1}) \in I_{m-1}$;

(5.2.2) $(x_1,\dots,x_m) \in I_m \Rightarrow (x_{\pi(1)},\dots,x_{\pi(m)}) \in I_m$

for every permutation π *of* $\{1,\dots,m\}$;

(5.2.3) $(x_1,\dots,x_m) \in I_m,\ (y_1,\dots,y_{m+1}) \in I_{m+1} \Rightarrow$

$\exists_{y\in\{y_1,\dots,y_{m+1}\}} [(x_1,\dots,x_m,y) \in I_{m+1}]$;

(5.2.4) $(x,x) \notin I_2$ *for every* $x \in S$.

If $\xi = (x_1,\dots,x_m) \in I_m$, then the sequence ξ is called *independent*, otherwise we say that ξ is *dependent*. Some simple properties follow immediately from this definition. First, it follows from (5.2.1) and (5.2.2) that a subsequence of an independent sequence is independent. Also, according to (5.2.4), a sequence is dependent whenever there are equal elements in it. We give two examples of independence relations.

Example 1. $(x_1,\dots,x_m) \in I_m$ if all x_k's are distinct.

Example 2. If S is a vector space, $(x_1,\dots,x_m) \in I_m$ if the vectors $x_1,\dots,x_m$ are linearly independent.

We need some further definitions for the formulation of the theorem.

DEFINITION 2. *A sequence* $(A_1,\dots,A_n)$ *of subsets of* S *is said to have property* H *(Rado's generalization of P. Hall's condition) if for each* $k \in \{1,\dots,n\}$ *the union of any* k*-tuple of* A_ν*'s contains* $x_1,\dots,x_k$ *such that* $(x_1,\dots,x_k) \in I_k$.

DEFINITION 3. *If* $(A_1,\dots,A_n)$ *is a sequence of subsets of* S, *then* $(x_1,\dots,x_n)$ *is called a system of independent representatives (abbreviated SIR) of* $(A_1,\dots,A_n)$ *if* $(x_1,\dots,x_n) \in I_n$ *and* $x_k \in A_k$ *(*$k = 1,\dots,n$*)*.

If a sequence $(A_1,\dots,A_n)$ has a SIR $(x_1,\dots,x_n)$, then $(A_1,\dots,A_n)$ has property H, since for every k-tuple $(A_{i_1},\dots,A_{i_k})$ we have

$$\{x_{i_1},\dots,x_{i_k}\} \subset A_{i_1} \cup \dots \cup A_{i_k} \quad \text{and} \quad (x_{i_1},\dots,x_{i_k}) \in I_k .$$

The theorem of Rado states the converse of this fact:

THEOREM 5.2.1 (Rado). *If a sequence* $(A_1,\dots,A_n)$ *of subsets of* S *has property* H, *then there exists a SIR of* $(A_1,\dots,A_n)$.

Specializing this theorem to example 1 we obtain P. Hall's theorem. In example 2, applying Rado's theorem gives the following result.

Corollary. Suppose that $A_1,\dots,A_n$ are subsets of a vector space, such that the union of any k-tuple of A_ν's contains at least k linearly independent vectors. Then there exist n linearly independent vectors $x_1,\dots,x_n$ with $x_k \in A_k$ $(k = 1,\dots,n)$.

This result is used in [8] for the derivation of stabilizability conditions for control systems.

The aim of this section is to give an algorithm for the computation of a SIR. The proof Rado gave for his theorem is very elegant, but it is not an algorithmic proof. It can be made algorithmic by a simple alteration, but then it turns out that the computation time T_n increases exponentially with n ($T_n \geq c.2^n$). It is possible however to shorten the algorithm such that (under "normal" conditions) the computation time grows algebraically with n (that is, $T_n = O(n^\lambda)$ for some positive λ). We assume in these considerations that the following basic operations can be performed:

B_1: "Given a subset $T \subset S$ and a number m, find out whether there exists a sequence $(x_1,\dots,x_m) \in I_m$ with $x_k \in T$ $(k = 1,\dots,m)$ and if so, find such a sequence."

B_2: "Given a sequence $(x_1,\dots,x_m)$, determine whether $(x_1,\dots,x_m) \in I_m$."

B_3: "If $(x_1,\dots,x_m) \in I_m$, $(y_1,\dots,y_{m+1}) \in I_{m+1}$, find $y \in \{y_1,\dots,y_{m+1}\}$ such that $(x_1,\dots,x_m,y) \in I_{m+1}$."

Operation B_2 is a special case of B_1, but since it is much simpler in general, it is stated separately. Operation B_3 is possible because of (5.2.3). It can be performed by a repeated application of B_2.

In the case of example 1 an algorithm for finding a SIR was given by M. Hall [9], but that algorithm does not readily extend to the general case.
We give Rado's proof of the theorem in the next part of the section and we indicate how the algorithm is derived from it. In the final part of this section we give the algorithm and an upper bound for the computation time.

Proof of Rado's theorem (see also [7]).

We proceed by induction. For n = 1 the result is clear. Suppose we have proved the result for n - 1. Let $(A_1,\dots,A_n)$ be a sequence of subsets of S satisfying *H*. Then also $(A_1,\dots,A_{n-1})$ has the property *H*. By the induction hypothesis there exists a SIR $(x_1,\dots,x_{n-1})$ of $(A_1,\dots,A_{n-1})$. According to *H* there exists $(y_1,\dots,y_n) \in I_n$ with $\{y_1,\dots,y_n\} \subset A_1 \cup \dots \cup A_n$. It follows from (5.2.3) that for some $y \in \{y_1,\dots,y_n\}$ we have $(x_1,\dots,x_{n-1},y) \in I_n$. If $y \in A_n$, we can put $y =: x_n$ and then $(x_1,\dots,x_n)$ is a SIR of $(A_1,\dots,A_n)$. Therefore we assume that $y \notin A_n$. Without loss of generality we may assume that $y \in A_1$. (If this is not the case we may interchange indices.) The collection $(A_2,\dots,A_n)$ satisfies property *H* and so, by induction hypothesis there exists a SIR $(v_2,\dots,v_n)$ of $(A_2,\dots,A_n)$. For every sequence $\varphi = (v_2,\dots,v_n)$ we define

(5. 2.5) $\rho(\varphi) := |\{k \mid x_k = v_k,\ k = 2,\dots,n-1\}|$,

where $|B|$ denotes the cardinality (that is, then number of elements) of B. Then we choose a SIR $\omega = (u_2,\dots,u_n)$ of $(A_2,\dots,A_n)$ such that $\rho(\omega)$ is maximal:

A_1	A_2	. . .	A_{n-1}	A_n
x_1	x_2	. . .	x_{n-1}	
y				
	u_2	. . .	u_{n-1}	u_n

Since $(x_1,\dots,x_{n-1},y) \in I_n$, there exists $z \in \{x_1,\dots,x_{n-1},y\}$ with $(z,u_2,\dots,u_n) \in I_n$. But, we cannot have $z = x_k$ for some $k \geq 2$. For, if this were the case, we would have $x_k \neq u_k$ and $\eta := (u_2,\dots,u_{k-1},x_k,u_{k+1},\dots,u_n) \in I_{n-1}$, according to the remarks following definition 1. That is, η would be a SIR of $(A_2,\dots,A_n)$ with $\rho(\eta) = \rho(\omega) + 1$, contradicting the definition of ω. Therefore, we have either $z = x_1$ or $z = y$. In both cases we put $u_1 := z$ and then $(u_1,u_2,\dots,u_n)$ is a SIR of $(A_1,A_2,\dots,A_n)$. This completes the proof.

This proof is not constructive in the sense that it manipulates an explicitly stated set of basic operations such as B_1, B_2, B_3. Actually if we take these operations to be the basic set, then it seems, in general, impossible to construct ω such that $\rho(\omega)$ is maximal. However, if instead of requiring that $\rho(\omega)$ be maximal we choose $\omega = (u_2,\dots,u_n)$ such that it is impossible to get a new SIR by replacing a $u_k \neq x_k$ with x_k, then the proof is still valid, but now ω can actually be constructed, once a SIR $(v_2,\dots,v_n)$ of $(A_2,\dots,A_n)$ is given. This can be done by replacing the v_k's successively with x_k's and checking whether the new sequences are independent. It is easily seen that this procedure stops after at most $\frac{1}{2}n(n-1)$ steps.

In the algorithm we thus obtain, the calculation of a SIR of length n requires the calculation of two SIR's of length $n-1$. Hence, if we denote by T_n the time needed for calculating a system of length n, then we have $T_n \geq 2T_{n-1}$ and therefore $T_n \geq 2^n \cdot c$ for some positive constant c. A closer consideration, however, makes clear that the algorithm can be shortened. In fact, for the calculation of the SIR $(v_2,\dots,v_n)$ of $(A_2,\dots,A_n)$ (before the maximization procedure) we need a SIR $(z_2,\dots,z_{n-1})$ of $(A_2,\dots,A_{n-1})$ and a SIR $(w_3,\dots,w_n)$ of $(A_3,\dots,A_n)$:

A_1	A_2	A_3	$\cdots$	A_{n-1}	A_n
	v_2	v_3	$\cdots$	v_{n-1}	v_n
	z_2	z_3	$\cdots$	z_{n-1}	
		w_3	$\cdots$	w_{n-1}	w_n
x_1	x_2	x_3	$\cdots$	x_{n-1}	

But since we have the SIR $(x_1,\ldots,x_{n-1})$ of $(A_1,\ldots,A_{n-1})$, we may just as well choose $(x_2,\ldots,x_{n-1})$ for $(z_2,\ldots,z_{n-1})$, and so it is not necessary to compute a new SIR of $(A_2,\ldots,A_{n-1})$. A repetition of this argument ultimately yields the algorithm given in the final part of this section.

The algorithm.

We suppose that we have constructed a SIR $(x_1,\ldots,x_{n-1})$ of $(A_1,\ldots,A_{n-1})$. The construction of a SIR $(u_1,\ldots,u_n)$ of $(A_1,\ldots,A_n)$ consists of two steps:

Step I: Construct a sequence $(y_1,\ldots,y_p)$ satisfying

(5.2.6) $$(y_k,x_k,\ldots,x_{n-1}) \in I_{n-k+1} \qquad (k = 1,\ldots,p)$$

(5.2.7) $$y_k \in A_k \quad (k = 1,\ldots,p-1)\ , \quad y_p \in A_n\ .$$

Step II: Construct a SIR of $(A_1,\ldots,A_n)$ with the aid of the elements $x_1,\ldots,x_{n-1}$, $y_1,\ldots,y_p$ obtained by Step I.

I. According to H, there exists $(z_1,\ldots,z_n) \in I_n$ with $z_k \in A_1 \cup \ldots \cup A_n$ $(k=1,\ldots,n)$. We can construct $(z_1,\ldots,z_n)$ by the operation B_1. Because of (5.2.3) (operation B_3) we can find $y_1 \in \{z_1,\ldots,z_n\}$ such that $(y_1,x_1,\ldots,x_{n-1}) \in I_n$. If $y_1 \in A_n$, we set $p=1$ and we start with step II. If $y_1 \notin A_n$, we change (if necessary) the numbering of the sets $(A_1,\ldots,A_{n-1})$ in such a way that $y_1 \in A_1$. Suppose we have already constructed $y_1,\ldots,y_{k-1}$, then because of H we can construct $(z_k,\ldots,z_n) \in I_{n-k+1}$ with $z_j \in A_k \cup \ldots \cup A_n$ $(j = k,\ldots,n)$. Again, owing to (5.2.3), we can find $y_k \in \{z_k,\ldots,z_n\}$ such that $(y_k,x_k,\ldots,x_{n-1}) \in I_{n-k+1}$. If $y_k \in A_n$, then we set $p := k$ and start with step II. Otherwise we change (if necessary) the numbering of $(A_k,\ldots,A_{n-1})$ so as to have $y_k \in A_k$, and we proceed with the construction of y_{k+1}. This procedure stops for some $k \leq n$, and it is possible that $p = n$.

A_1	$\cdots$	A_{p-1}	A_p	$\cdots$	A_{n-1}	A_n
x_1	$\cdots$	x_{p-1}	x_p	$\cdots$	x_{n-1}	
y_1	$\cdots$	y_{p-1}				y_p

II. We can construct a SIR $(u_1,\dots,u_n)$ as follows. We set $u_n := y_p$ and $u_k := x_k$ for $k = n-1,n-2,\dots,p$ (if $p < n$). For the construction of u_k for $k = p-1,\dots,1$ we assume that u_{k+1} has already been constructed, and we set $u_k := x_k$ if $(x_k,u_{k+1},\dots,u_n) \in$ $\in I_{n-k+1}$ and $u_k := y_k$ otherwise.

Let us show that the sequence $(u_1,\dots,u_n)$, thus constructed, is a SIR of $(A_1,\dots,A_n)$. Obviously, we have $u_k \in A_k$ $(k = 1,\dots,n)$, so it is sufficient to prove that $(u_1,\dots,u_n)$ is independent. We proceed by induction. Because of (5.2.6) we have that $(u_p,\dots,u_n) = (x_p,\dots,x_{n-1},y_p)$ is independent. If for some k with $1 \le k < p$ we have shown that $(u_{k+1},\dots,u_n)$ is independent, then, in order to prove that $(u_k,\dots,u_n)$ is independent, we may assume that $u_k = y_k$ by the construction of u_k (see II). Because of (5.2.6) we know that $(y_k,x_k,\dots,x_{n-1}) \in I_{n-k+1}$. Hence, according to (5.2.3) there exists $z \in \{y_k,x_k,\dots,x_{n-1}\}$ with

$$(5.2.8) \qquad (z,u_{k+1},\dots,u_n) \in I_{n-k+1} .$$

We show that we must have $z = y_k$. In fact, if $z = x_\nu$ for some $\nu \in \{k,\dots,n-1\}$, then $u_\nu \neq x_\nu$, since otherwise $(z,u_{k+1},\dots,u_\nu,\dots,u_n)$ would be dependent. Hence, $u_\nu = y_\nu$. But then according to the construction of u_ν we must have that $(x_\nu,u_{\nu+1},\dots,u_n)$ is dependent. Since $\nu \ge k$ and $z = x_\nu$, this is in contradiction with (5.2.8). Therefore $z = y_k = u_k$, and hence $(u_k,\dots,u_n)$ is independent.

Remark 1. It sould be emphasized that step II and the proof following it also apply if $p = n$, although some parts of it become irrelevant in that case. Furthermore, the alteration of the numbering of $(A_1,\dots,A_n)$ in step I is not necessary. It is only performed in order to facilitate both the formulation of step II, and the proof that $(u_1,\dots,u_n)$ is a SIR of $(A_1,\dots,A_n)$.

We give an estimate for the time T_n needed for the calculation of a SIR of $(A_1,\dots,A_n)$. We assume that there exist real numbers $K_1 > 0$, $\alpha > 0$, such that for all finite sets $T \subset S$ operation B_1 requires a computation time $\le K_1|T|^\alpha$. Also, we assume that there exists $K_2,K_3 > 0$, $\beta,\gamma > 0$ such that operations B_2 and B_3 require a computation time $\le K_2m^\beta$ and $\le K_3m^\gamma$, respectively. It is easily verified that this is satisfied in examples 1 and 2. Furthermore, we make the assumption that the sets $A_1,\dots,A_n$ have cardinality of order n^ν. We see that step I requires operations B_1 and B_3 at most n times and step II requires operation B_2 at most n times. For operation B_1 we always have $T = A_{i_1} \cup \dots \cup A_{i_k}$ and hence $|T| = O(n^{\nu+1})$. So, if $\lambda := \max(\alpha(\nu+1),\beta,\gamma)$, then $T_n \le T_{n-1} + Cn^{\lambda+1}$ and hence $T_n = O(n^{\lambda+2})$.

Remark 2. In the case of example 1, step II can be simplified. Indeed, it is not hard to show that $(u_{k+1},\dots,u_n)$ contains exactly one element w_k not in $(x_{k+1},\dots,x_{n-1})$, and that $u_k = y_k$ if and only if $x_k = w_k$. So we can replace step II in the algorithm by

II': Set $w_p := u_n := y_p$, $u_k := x_k$ $(k = n-1,\dots,p)$. Suppose that $1 \le k < p$ and that u_{k+1}, w_{k+1} have been constructed, then define u_k, w_k by:
If $x_k \neq w_{k+1}$ then $u_k := x_k$, $w_k := w_{k+1}$;
if $x_k = w_{k+1}$ then $w_k := u_k := y_k$.

So in this case we only have to compare x_k with w_k in order to decide whether u_k should be x_k or y_k. It also follows from this observation that we have $\{u_1,\dots,u_n\} \supset \{x_1,\dots,x_{n-1}\}$. Note that this is also the case in M. Hall's algorithm (cf. [9]) although the algorithms seem to be quite different.

Remark 3. As is observed in Remark 2, we have in the case of example 1, that the set $\{u_{k+1},\dots,u_n\} \setminus \{x_{k+1},\dots,x_{n-1}\}$ contains exactly one element w_k. It is possible to extend this result to the general case. To that aim, we introduce a new concept:

DEFINITION 4. *If* $(x_1,\dots,x_n) \in I_n$, *then the span of* $x_1,\dots,x_n$ *is defined to be the set*

$$\langle x_1,\dots,x_n\rangle := \{x \in S \mid (x_1,\dots,x_n,x) \notin I_{n+1}\} .$$

Obviously $x_k \in \langle x_1,\dots,x_n\rangle$ for $k = 1,\dots,n$. We mention two further properties of the span.

Proposition 1. If $u_1,\dots,u_{n+1} \in \langle x_1,\dots,x_n\rangle$, then $(u_1,\dots,u_{n+1})$ is dependent.

Proof. This follows immediately from (5.2.3).

Proposition 2. If $u_1,\dots,u_n \in \langle x_1,\dots,x_n\rangle$ and $(u_1,\dots,u_n)$ is independent, then $\langle u_1,\dots,u_n\rangle = \langle x_1,\dots,x_n\rangle$.

Proof. If $y \notin \langle u_1,\dots,u_n\rangle$, then $(u_1,\dots,u_n,y) \in I_{n+1}$. By (5.2.3) there exists $z \in \{u_1,\dots,u_n,y\}$ with $(x_1,\dots,x_n,z) \in I_{n+1}$. Since $u_k \in \langle x_1,\dots,x_n\rangle$ for $k = 1,\dots,n$, we must have $z = y$, that is, $y \notin \langle x_1,\dots,x_n\rangle$. Thus we have shown that $\langle x_1,\dots,x_n\rangle \subset \langle u_1,\dots,u_n\rangle$. It follows that $x_k \in \langle x_1,\dots,x_n\rangle \subset \langle u_1,\dots,u_n\rangle$. Therefore we may interchange the role of $u_1,\dots,u_n$ and $x_1,\dots,x_n$ in the foregoing argument. Then we obtain $\langle u_1,\dots,u_n\rangle \subset \langle x_1,\dots,x_n\rangle$.

We have the following result:

THEOREM 5.2.2. *Let* $(A_1,\dots,A_n)$ *satisfy condition H and let* $(x_1,\dots,x_{n-1})$ *be a SIR of* $(A_1,\dots,A_{n-1})$. *Then the SIR* $(u_1,\dots,u_n)$ *of* $(A_1,\dots,A_n)$ *constructed in the algorithm has the property that* $(u_k,\dots,u_n)$ *contains exactly one element* $w_k \notin \langle x_k,\dots,x_{n-1}\rangle$ *for* $k = 1,\dots,n-1$. *In particular,* $(u_1,\dots,u_n)$ *contains exactly one element not contained in* $\langle x_1,\ldots,x_{n-1}\rangle$.

Proof. We proceed by backward induction. The result is clear for $k = n$, since $y \in \langle\emptyset\rangle$ implies $(y) \notin I_1$ whereas $(y_p) \in I_1$. If the property has been proved for $k+1$, then according to step II there are two possibilities:

(i) $(x_k,u_{k+1},\ldots,u_n) \in I_{n-k+1}$.

Then $u_k = x_k \in \langle x_k,\ldots,x_n\rangle$, so that $w_{k-1} = w_k$.

(ii) $(x_k,u_{k+1},\ldots,u_n) \notin I_{n-k+1}$.

Then $u_k = y_k$. Let $(v_{k+1},\ldots,v_{n-1})$ denote the sequence $(u_{k+1},\ldots,u_n)$ with the element w_k deleted. Then $(v_{k+1},\ldots,v_{n-1})$ is independent. Also, by induction hypothesis, $v_{k+1},\ldots,v_{n-1} \in \langle x_{k+1},\ldots,x_{n-1}\rangle$. It follows from Proposition 2, that $\langle x_{k+1},\ldots,x_{n-1}\rangle = \langle v_{k+1},\ldots,v_{n-1}\rangle$. Therefore, we have $x_k \notin \langle v_{k+1},\ldots,v_{n-1}\rangle$, that is, $(x_k,v_{k+1},\ldots,v_{n-1}) \in I_{n-k}$. Since $(x_k,u_{k+1},\ldots,u_n)$ and hence $(x_k,w_k,v_{k+1},\ldots,v_{n-1})$ is dependent, we conclude that $w_k \in \langle x_k,v_{k+1},\ldots,v_{n-1}\rangle = \langle x_k,x_{k+1},\ldots,x_{n-1}\rangle$. We have shown that $u_{k+1},\ldots,u_{n-1} \in \langle x_k,\ldots,x_{n-1}\rangle$. The proof is complete if we show that $u_k = y_k$ is not in $\langle x_k,\ldots,x_{n-1}\rangle$, so that $w_{k-1} = u_k$. However, this follows immediately from Proposition 1, because $(y_k,u_{k+1},\ldots,u_{n-1})$ is independent and $u_{k+1},\ldots,u_{n-1} \in \langle x_k,\ldots,x_{n-1}\rangle$.

In the case of example 1, $\langle x_1,\ldots,x_n\rangle = \{x_1,\ldots,x_n\}$ so that the theorem is a generalization of the property mentioned in Remark 2. It should be noticed that the more obvious generalization $|\{u_k,\ldots,u_n\} \setminus \{x_k,\ldots,x_{n-1}\}| = 1$ is not correct in the general case. A counterexample can be given for the case of example 2. Let a_1, a_2 be independent vectors. $A_1 := \{a_1,a_2\}$, $A_2 = \{2a_1\}$. Then (a_1) is a SIR of (A_1). The only SIR of (A_1,A_2) is $(a_2,2a_1)$ which does not contain a_1.

Finally we remark that step I of the algorithm may be modified. For instance, instead of choosing the sequence $(z_1,\ldots,z_n)$ and the element $y_1 \in \{z_1,\ldots,z_n\}$ arbitrarily, one may try several choices trying to obtain that $y_1 \in A_n$. Similar remarks apply to the construction of y_k for $k \geq 2$. In that way one possibly finds a smaller p.

5.3. *A problem on SDR's.*

The following theorem was proposed as a problem by V. Chvátal [10]. The proof presented here is due to J.H. Timmermans.

THEOREM 5.3.1. *Let* $A_1,A_2,\ldots,A_n$ *be finite sets. If*

$$\sum_{1\leq i<j\leq n} \frac{|A_i \cap A_j|}{|A_i|\cdot|A_j|} < 1$$

then the sets $A_1,\dots,A_n$ *have a system of distinct representatives.*

Proof. Let $T := (A_1,A_2,\dots,A_n)$ be a sequence of finite sets which has no system of distinct representatives and let

$$S_T = \sum_{1\le i<j\le n} \frac{|A_i \cap A_j|}{|A_i|\cdot|A_j|} .$$

Let $1 \le \nu \le n$. If $A_\nu = \{x_1,x_2,\dots,x_\ell\}$ we define $A_{\nu\mu} := \{x_\mu\}$.
Let T_μ be the sequence obtained from T by replacing A_ν by $A_{\nu\mu}$. Clearly T_μ does not have a system of distinct representatives. We have

$$S_{T_\mu} = S_T - \frac{1}{\ell} \sum_{j\ne\nu} \frac{|A_\nu \cap A_j|}{|A_j|} + \sum_{j\ne\nu} \frac{|A_{\nu\mu} \cap A_j|}{|A_j|} .$$

Since

$$\sum_{\mu=1}^{\ell} |A_{\nu\mu} \cap A_j| = |A_\nu \cap A_j|$$

we find

$$\sum_{\mu=1}^{\ell} S_{T_\mu} = \ell S_T .$$

It follows that μ can be chosen in such a way that $S_{T_\mu} \le S_T$. We can do this for $\nu = 1,2,\dots,n$. In this way we find a sequence $T' = (A_1',A_2',\dots,A_n')$ with $|A_i'| = 1$ $(i = 1,2,\dots,n)$ such that T' does not have a system of distinct representatives and $S_{T'} \le S_T$. Now two of the sets A_i' must be identical, say $A_i' = A_j'$, and then

$$S_{T'} \ge \frac{|A_i' \cap A_j'|}{|A_i'|\cdot|A_j'|} = 1 .$$

This proves the assertion.

Remark 1. The example $A_i := \{1,i+1\}$, $i = 1,2,\dots,n$, shows that the condition of Theorem 5.3.1 is not necessary.

Remark 2. The proof given by Chvátal (cf. [10]) gives a lower bound for the number of SDR's of the sets $A_1,A_2,\dots,A_n$. This bound is often poor. Representatives of $A_1,A_2,\dots,A_n$ can be chosen in $|A_1|\cdot|A_2|\cdot\ldots\cdot|A_n|$ ways. If we require the representatives of A_i and A_j to be the same $(i \ne j)$, then this can be done in $|A_1|\cdot\ldots\cdot|A_n|\cdot \frac{|A_i \cap A_j|}{|A_i|\cdot|A_j|}$ ways. Therefore the number of SDR's is at least

$$(5.3.1) \qquad |A_1|\cdot\ldots\cdot|A_n| \,\{1 - \sum_{1\le i<j\le n} \frac{|A_i \cap A_j|}{|A_i|\cdot|A_j|}\} .$$

5.4. *An application of linear recurrences.*

We consider a generalization of the first problem on H.p.53. The sets $S_1, S_2, \ldots, S_n$ are the columns of the array (5.4.1)

$$(5.4.1)\qquad \begin{matrix} 1 & 2 & 3 & \cdots & n-2 & n-1 & n \\ 2 & 3 & 4 & \cdots & n-1 & n & 1 \\ 4 & 5 & 6 & \cdots & 1 & 2 & 3 \end{matrix}$$

The problem is to determine the number of SDR's of these sets. This problem is strongly connected to the material of H. § 3.2. We first consider two problems of the same type as those treated in H. § 3.2 and apply the results to our problem. Consider the array (5.4.2)

$$(5.4.2)\qquad \begin{matrix} & 1 & 2 & 3 & \cdots & t-3 & t-2 & t-1 \\ 1 & 2 & 3 & 4 & \cdots & t-2 & t-1 & t \\ 3 & 4 & 5 & 6 & \cdots & t & & \end{matrix}$$

Let u_t, $t \geq 3$, be the number of ways of finding a permutation $a_1, a_2, \ldots, a_t$ of $1,2,\ldots,t$ such that for each i, a_i is in the i-th column of the array (5.4.2). There are two possibilities: either the permutation has $a_t = t$ or it has $(a_{t-2}, a_{t-1}, a_t) = (t, t-2, t-1)$. This yields the recurrence

$$(5.4.3)\qquad u_t = u_{t-1} + u_{t-3} .$$

If we extend the definition of u_t by taking $u_0 = u_1 = u_2 = 1$ the sequence u_t starts with $1,1,1,2,3,4,6,9,\ldots$. By standard methods we find

$$(5.4.4)\qquad u_t = \sum_{i=1}^{3} \alpha_i (x_i)^t ,$$

where x_1, x_2, x_3 are the zeros of $x^3 - x^2 - 1$.

Next we consider in the same way the array (5.4.5):

$$(5.4.5)\qquad \begin{matrix} & & 1 & 2 & 3 & \cdots & t-3 & t-2 \\ & 1 & 2 & 3 & 4 & \cdots & t-2 & t-1 \\ 2 & 3 & 4 & 5 & 6 & \cdots & t & \end{matrix}$$

Here, let v_t be the number of permutations of $1,2,\ldots,t$ which are a SDR of the columns of (5.4.5) where we have now formulated the analogous problem in the terminology of this chapter. We find $v_0 = 0$, $v_1 = v_2 = v_3 = v_4 = 1$, $v_5 = 2$, and

$$(5.4.6)\qquad v_t = v_{t-2} + v_{t-3} \qquad (t \geq 5) ,$$

i.e. v_t is a linear combination of the t-th powers of the zeros of $x^3 - x - 1$.

Now we turn to the difficult problem of finding the number A_n of SDR's of the sets in the array (5.4.1). This is done by choosing representatives for the final three columns and then reducing the problem to one of the previous problems. We treat one case in detail as an example. Suppose we choose $n-2$, 2, n as representatives of the last

3 columns in (5.4.1). We show this in (5.4.7) by circling the chosen representatives.

(5.4.7)

(1)	2	3	4	...	n-6	n-5	n-4	(n-3)	(n-2)	n-1	(n)
2	3	4	5	...	n-5	n-4	n-3	n-2	n-1	n	1
4	5	6	7	...	n-3	n-2	(n-1)	n	1	(2)	3

We have also circled the numbers 1 in column 1, n-3 in column n-3 and n-1 in column n-4 because there is no other possible choice. The remaining representatives must be chosen from the columns of the array (5.4.8):

(5.4.8)

	3	4	...	n-7	n-6	n-5
3	4	5	...	n-6	n-5	n-4
5	6	7	...	n-4		

and this is possible in u_{n-6} ways since (5.4.2) and (5.4.8) are similar. Below we list the possible choices for the representatives of the last three columns in (5.4.1) and for each of these the number of ways of completing the SDR. The reader should have no trouble checking the details in the manner demonstrated above.

Representatives	# of SDR's	Representatives	# of SDR's
n-2, n-1, n	1	n-2, n, 3	u_{n-7}
1, n-1, n	u_{n-3}	n-2, 2, 1 }	v_{n-3}
n-2, 2, n	u_{n-6}	n-2, 2, 3 }	
n-2, n-1, 1	u_{n-4}	n-1, n, 1	u_{n-3}
n-2, n-1, 3	u_{n-6}	n-1, 2, 1	v_{n-4}
n-1, 2, n	u_{n-4}	n-1, n, 3	u_{n-5}
1, 2, n	v_{n-3}	1, n, 3	v_{n-2}
1, n-1, 3	v_{n-3}	n-1, 2, 3	v_{n-5}
n-2, n, 1	u_{n-5}	1, 2, 3	1

Adding all possibilities and applying (5.4.3) and (5.4.6) we find

(5.4.9) $$A_n = (u_n + 2u_{n-3}) + (2v_n + v_{n-3}) + 2 \qquad (n \geq 6) .$$

For n = 4,5,6,7,8,... this yields A_n = 9,13,17,24,33,...
From (5.4.9) we see that it would have been extremely difficult to find a recurrence for A_n directly since this would involve 7 terms.

5.5. *Permanents*.

In this chapter we have already remarked that, if $S_1, S_2, \dots, S_n$ are subsets of $\{1,2,\dots,n\}$ and if the (0,1) matrix A is defined by $A := [a_{ij}]$ with

$$a_{ij} := \begin{cases} 1 & \text{if } j \in S_i, \\ 0 & \text{otherwise} \end{cases}$$

then per(A) is the number of SDR's of the system $S_1, S_2, \dots, S_n$. In the corollary to Theorem 5.1.2 this led to the bound

$$\text{per}(A) \geq \prod_{i=1}^{n} (r_i - i + 1)_* \tag{5.5.1}$$

if the rowsums r_i are ordered such that they satisfy $r_1 \leq r_2 \leq \dots \leq r_n$ and if we also assume that per(A) ≠ 0. This is better than the lower bounds previously known (cf. [6]) but the restriction that we must assume per(A) ≠ 0 is unpleasant.
It was shown by H. Minc [11] that

$$\text{per}(A) \leq \prod_{i=1}^{n} \left(\frac{r_i + \sqrt{2}}{1 + \sqrt{2}}\right). \tag{5.5.2}$$

As an example we now consider all n × n (0,1) matrices A in which all the rowsums and columnsums are 2. A special example is $I + P_n$ where P_n is the permutation matrix with $p_{ij} = 1$ iff $j - i \equiv 1 \pmod n$. It is obvious that $\text{per}(I + P_n) = 2$. This shows that the inequality (5.5.1) is sharp in this case. On the other hand, any (0,1) matrix which has two 1's in every row and column is equivalent to a direct sum of matrices of type $(I + P_n)$. Therefore we have

THEOREM 5.5.1. *If* A *is a* (0,1) *matrix in which all rowsums and columnsums are* 2 *then*

$$\text{per } A \leq 2^{[\frac{1}{2}n]}. \tag{5.5.3}$$

Equality is possible in (5.5.3) if n is even and A is the direct sum of $\frac{1}{2}n$ 2 by 2 matrices J. In this case we also have equality in the upper bound (5.5.2).
Now let us consider n by n matrices A in which all rowsums and columnsums are k. If n is a multiple of k then we can take A to be the direct sum of $\frac{n}{k}$ matrices J of size k. Then $\text{per}(A) = (k!)^{n/k}$. It was conjectured by H.J. Ryser that this is the maximal permanent in this class. This is a special case of the following conjecture of H. Minc ([12]):

CONJECTURE. *If* A *is a* (0,1) *matrix of order* n *with rowsums* $r_1, r_2, \dots, r_n$, *then*

$$\text{per}(A) \leq \prod_{j=1}^{n} (r_j!)^{1/r_j}. \tag{5.5.4}$$

A result comparable to (5.5.4) and (5.5.2) is due to W.B. Jurkat and H.J. Ryser [13]:

THEOREM 5.5.2. *If* A *is a* (0,1) *matrix of order* n *with rowsums* $r_1, r_2, \ldots, r_n$, *then*

$$(5.5.5) \qquad \mathrm{per}(A) \le \prod_{j=1}^{n} (r_j!)^{1/n} \left(\frac{r_j + 1}{2}\right)^{(n-r_j)/n} .$$

We now take a closer look at the upper bounds (5.5.2) and (5.5.4). Let φ be a function defined on $\mathbb{N}$ and consider a theorem of the type of (5.5.2) and (5.5.4), namely

$$(5.5.6) \qquad \mathrm{per}(A) \le \prod_{j=1}^{n} \varphi(r_j)$$

which we now wish to prove by induction on n. Clearly, it is necessary that $\varphi(1) \ge 1$ and from (5.5.3) it follows that $\varphi(2) \ge 2^{\frac{1}{2}}$ is also a necessary condition on such a function φ. Note that $\varphi_1(n) := (n!)^{1/n}$ and $\varphi_2(n) := \frac{n + \sqrt{2}}{1 + \sqrt{2}}$ (the functions occurring in (5.5.4) and (5.5.2)) satisfy these inequalities with equality holding in both cases. To prove (5.5.6) by induction we note that the theorem is trivial for n = 1 and n = 2 and then assume it holds for some n. An n + 1 by n + 1 matrix is considered and we permute the rows in such a way that the 1's in columns 1 are in the first ℓ places. In the proof we may always assume no rowsum or columnsum is 1 (this being the trivial case). Developing by the first column we find that we are done if the following inequality holds:

$$(5.5.7) \qquad \sum_{i=1}^{\ell} \frac{1}{\varphi(r_i - 1)} \le \prod_{i=1}^{\ell} \frac{\varphi(r_i)}{\varphi(r_i - 1)} \qquad (\text{all } r_i \ge 2) .$$

We first remark that it is easily checked that both φ_1 and φ_2 satisfy (5.5.7) for $\ell = 2$. Now (5.5.7) may possibly be proved by induction on ℓ or on the r_i's or both. We assume $\varphi(1) = 1$. Then, taking all $r_i = 2$, (5.5.7) becomes

$$\ell \le \{\varphi(2)\}^{\ell} ,$$

and hence

$$\varphi(2) \ge \max\{\ell^{1/\ell} \mid \ell \in \mathbb{N}\} = 3^{1/3} .$$

Now assume (5.5.7) holds for some $\ell \ge 2$ and all ℓ-tuples $(r_1, r_2, \ldots, r_\ell)$ $(r_i \ge 2,$ $i = 1,2,\ldots,\ell)$. Consider $\ell + 1$. If $r_1 = r_2 = \ldots = r_{\ell+1} = 2$ then again (5.5.7) holds. We write (5.5.7) as follows

$$(5.5.8) \qquad \sum_{i=1}^{\ell} \frac{\varphi(r_1 - 1)\varphi(r_2 - 1) \ldots \varphi(r_\ell - 1)}{\varphi(r_i - 1)} \le \varphi(r_1)\varphi(r_2) \ldots \varphi(r_\ell) .$$

If we consider (5.5.8) for $\ell + 1$ instead of ℓ and replace $r_{\ell+1}$ by $r_{\ell+1} + 1$, then the left-hand side is increased by

$$\{\varphi(r_{\ell+1}) - \varphi(r_{\ell+1} - 1)\} \sum_{i=1}^{\ell} \frac{\varphi(r_1 - 1) \ldots \varphi(r_\ell - 1)}{\varphi(r_i - 1)}$$

and the right-hand side is increased by

$$\{\varphi(r_{\ell+1} + 1) - \varphi(r_{\ell+1})\}\varphi(r_1)\varphi(r_2) \ldots \varphi(r_\ell) .$$

Apparently a proof by induction of (5.5.7) and hence of (5.5.6) presents no problems if φ is convex. The best possible theorem we can prove in this way is obtained by taking for φ the linear function with $\varphi(1) = 1$, $\varphi(2) = 3^{1/3}$.

THEOREM 5.5.3. *If* A *is a* (0,1) *matrix of order* n *with rowsums* $r_1, r_2, \ldots, r_n$, *then*

(5.5.9) $$\mathrm{per}(A) \le \prod_{j=1}^{n} \{(3^{1/3} - 1)r_i + 2 - 3^{1/3}\} .$$

The best possible theorem one can hope for using this method of proof is (5.5.6) with for φ the linear function with $\varphi(1) = 1$ and $\varphi(2) = 2^{\frac{1}{2}}$. But then $\varphi = \varphi_2$ and we have Minc's theorem (5.5.2). Since in this case (5.5.7) is not true for all ℓ-tuples $(r_1, r_2, \ldots, r_\ell)$ the proof has to be modified. In [11] Minc does this as follows. If $A = [a_{ij}]$ has rowsums r_i $(i = 1,2,\ldots,n)$ and $b_{ij} := a_{ij}/r_i$, then B has rowsums 1. Therefore B has a column with sum ≤ 1 (in fact, even a column with sum < 1 unless all columnsums of B are 1). We permute columns of A such that the first column has this property, i.e.

(5.5.10) $$\sum_{i=1}^{\ell} r_i^{-1} \le 1 .$$

Minc then proceeds to proof (5.5.7) under the condition (5.5.10) directly instead of by induction. Since φ_1 is not convex we cannot expect this idea to work in that case. We shall now show that even the idea of proving (5.5.6) by induction is not promising in that case. If $\varphi = \varphi_1$ then we have

$$\frac{\varphi_1(r_i)}{\varphi_1(r_i - 1)} = \left(\frac{r_i}{\varphi_1(r_i - 1)}\right)^{1/r_i}$$

and therefore we can rewrite (5.5.7) as follows

(5.5.11) $$\sum_{i=1}^{\ell} \frac{1}{r_i} \frac{r_i}{\varphi_1(r_i - 1)} \le \prod_{i=1}^{\ell} \left(\frac{r_i}{\varphi_1(r_i - 1)}\right)^{1/r_i} .$$

By the arithmetic-geometric mean inequality, (5.5.11) is false if $\sum_{i=1}^{\ell} r_i^{-1} = 1$ unless all the r_i's are equal! It is easy to find a (0,1) matrix for which $\sum_{i=1}^{\ell} r_i^{-1} = 1$ for every column (without using equal r_i's), e.g.

$$(5.5.12) \qquad \begin{pmatrix} 1 & 1 & 0 & 0 \\ 1 & 1 & 1 & 1 \\ 1 & 1 & 1 & 1 \\ 0 & 0 & 1 & 1 \end{pmatrix}$$

In [12] Minc remarks that his conjecture (5.5.4) has been checked (for all n) if all the rowsums are ≤ 5. He did not publish this proof. However, from what was proved above it is easily seen how such a proof goes. First assume all rowsums are ≤ 3. We use induction. In (5.5.7) we may assume $\sum_{i=1}^{\ell} r_i^{-1} \le 1$ and by the remark following (5.5.7) we may take $\ell \ge 3$. This is possible only if $\ell = 3$, $r_1 = r_2 = r_3 = 3$ and then equality holds in (5.5.7). Now assume all rowsums are ≤ 4. Again $\ell \ge 3$ and we now may assume that at least one of the r_i's is 4. For $(r_1, r_2, \ldots, r_\ell)$ this leaves the possibilities (4,4,4,4), (4,4,4), (3,4,4), (3,3,4), (2,4,4). Now the difficulty mentioned above arises. The inequality (5.5.7) is true for each of these possibilities except (2,4,4) where $\sum r_i^{-1} = 1$. Hence we cannot use (5.5.7) in the case where all the columns of A are of this type (cf. (5.5.12)). Assume (5.5.4) is true for all matrices of size $< n$ if all $r_i \le 4$. Let A be of size n and let A in the standard form used above have $\ell = 3$, $r_1 = 2$, $r_2 = r_3 = 4$. Develope per(A) by the first row. We find, using (5.5.4) and the fact that the second column also has sum 3 corresponding to rows with sum 2, 4 and 4,

$$\mathrm{per}(A) \le 2\left(\frac{\varphi_1(3)}{\varphi_1(4)}\right)^2 \prod_{i=2}^{n} \varphi_1(r_i) < \varphi_1(2) \prod_{i=2}^{n} \varphi_1(r_i) = \prod_{i=1}^{n} \varphi_1(r_i) \,.$$

This proves (5.5.4) under the condition that all rowsums are ≤ 4. It does not seem worthwhile to continue this line of attack.

A theorem very close to Minc's conjecture (5.5.4) was given by A. Nijenhuis and H.S. Wilf [16]. They define the function φ_3 recursively by

$$(5.5.13) \qquad \begin{aligned} &\varphi_3(1) = 1 \,, \\ &\varphi_3(n+1) = \varphi_3(n) \exp(1/(e\varphi_3(n))) \,. \end{aligned}$$

Then, by substitution we find (using the notation $x := \sum_{i=1}^{\ell} \frac{1}{\varphi_3(r_i - 1)}$):

$$\sum_{i=1}^{\ell} \frac{1}{\varphi_3(r_i - 1)} \prod_{j=1}^{\ell} \frac{\varphi_3(r_i - 1)}{\varphi_3(r_i)} = \sum_{i=1}^{\ell} \frac{1}{\varphi_3(r_i - 1)} \prod_{j=1}^{\ell} \exp\left(-\frac{1}{e\varphi_3(r_j - 1)}\right) = xe^{-x/e} \,.$$

Since $xe^{-x/e} \le 1$ for $x > 0$ we see that φ_3 satisfies the inequality (5.5.7), i.e. the inequality (5.5.6) is true for $\varphi = \varphi_3$. In [16] it is shown that

$$(5.5.14) \qquad \varphi_3(n) = \frac{n}{e} + \frac{\log n}{2e} + \frac{A}{e} + o(1) \qquad (n \to \infty) \,.$$

Since

$$\varphi_1(n) = (n!)^{1/n} = \frac{n}{e} + \frac{\log n}{2e} + C + o(1) \qquad (n \to \infty)$$

the following theorem holds:

THEOREM 5.5.4. *There is a constant* c *such that for every* (0,1) *matrix* A *of order* n *with rowsums* $r_1, r_2, \ldots, r_n$ *we have*

$$\text{(5.5.15)} \qquad \operatorname{per}(A) \le \prod_{j=1}^{n} \{(r_j!)^{1/r_j} + c\} .$$

The improvement compared to (5.5.2) is that the first two terms of the asymptotic expansion of $\varphi_3(n)$ correspond to those of $\varphi_1(n)$, which is not even true for the first term for $\varphi_2(n)$. If one is concerned in this fact only, and not in a good estimate of c, then a shorter proof than in [16] can be given by considering the function φ_4 defined as follows:

$$e\varphi_4(n) := 1 + n + \tfrac{1}{2} \sum_{k=1}^{n} k^{-1} = n + \tfrac{1}{2}\log n + B + o(1) \qquad (n \to \infty) .$$

From

$$x(e^{1/x} - 1) = 1 + \sum_{n=2}^{\infty} \frac{x^{-n+1}}{n!} \le 1 + \tfrac{1}{2}(x - \tfrac{1}{3})^{-1}$$

we find, by substituting $x = e\varphi_4(n) \ge n + \frac{3}{2}$,

$$\varphi_4(n+1) \ge \varphi_4(n) \exp\left(\frac{1}{e\varphi_4(n)}\right) ,$$

i.e. (5.5.7) and (5.5.6) also hold for φ_4 (by the same reasoning we used for φ_3). Of course this last theorem is not good if all the rowsums are small, since we already know that (5.5.4) is true and best possible if all $r_i \le 4$.

We now turn back to the problem we are especially interested in, namely (0,1) matrices for which all rowsums and columnsums are k. This case is connected with the famous Van der Waerden conjecture:

CONJECTURE. *If* A *is a matrix with nonnegative entries in which all rowsums and columnsums are* 1, *then*

$$\text{(5.5.16)} \qquad \operatorname{per}(A) \ge n!\, n^{-n} .$$

Let $\mathfrak{A}(n,k)$ be the class of (0,1) matrices of order n with k ones in each row and column. We define

$$(5.5.17) \qquad M(n,k) := \mathrm{Max}\{\mathrm{per}(A) \mid A \in \mathfrak{A}(n,k)\} ,$$

$$(5.5.18) \qquad m(n,k) := \mathrm{Min}\{\mathrm{per}(A) \mid A \in \mathfrak{A}(n,k)\} .$$

By taking direct sums it is immediately clear that

$$(5.5.19) \qquad M(n_1+n_2,k) \ge M(n_1,k)M(n_2,k) ,$$

$$(5.5.20) \qquad m(n_1+n_2,k) \le m(n_1,k)m(n_2,k) .$$

From this it follows that

$$(5.5.21) \qquad M(k) := \lim_{n\to\infty} \{M(n,k)\}^{1/n}$$

and

$$(5.5.22) \qquad m(k) := \lim_{n\to\infty} \{m(n,k)\}^{1/n}$$

exist.

We can now formulate the results of the previous pages in terms of $M(k)$ and $m(k)$. From (5.5.5) we find that $M(k) \le \frac{k+1}{2}$ which is not as good as the following consequence of (5.5.2):

$$(5.5.23) \qquad M(k) \le (\sqrt{2} - 1)k + (2 - \sqrt{2}) .$$

Our example shows that

$$(5.5.24) \qquad M(k) \ge (k!)^{1/k}$$

and conjecture (5.5.4) says that equality holds in (5.5.24).

By Theorem 5.5.1 we have $M(2) = 2^{\frac{1}{2}}$. We have proved equality in (5.5.24) for $k = 3$ and for $k = 4$ above. Clearly $m(2) = 1$. From the Van der Waerden conjecture we would get $m(k) \ge \frac{k}{e}$ but nothing even close to this has been proved. In fact the best estimates at the moment are ([14], [15]):

$$(5.5.25) \qquad m(n,k) \ge \frac{(k-1)(k-2)}{2} n$$

$$(5.5.26) \qquad m(n,3) \ge n + 3 .$$

This only gives the trivial result $m(k) \ge 1$. Note that the Van der Waerden conjecture and Minc's conjecture (5.5.4) imply that $\lim_{k\to\infty} m(k)/M(k) = 1$.

By considering special classes of (0,1) matrices one can obtain bounds for $M(k)$ and $m(k)$ as we have shown for $M(k)$. In fact, the result of Section 5.4 gives an upper bound for $m(3)$, namely

$$(5.5.27) \qquad m(3) \le \xi \approx 1{,}465 ,$$

where ξ is the largest zero of $x^3 - x^2 - 1$. The class which gives this bound is $I + P_n + P_n^3$. If we treat the class $I + P_n + P_n^4$ in the same way as in Section 5.4 we obtain a slight improvement of (5.5.27). One is tempted to expect the permanent of $I + P_n + P_n^{\ell}$ to be small if ℓ is about $\frac{1}{2}n$. However, this is not true. Take $n = 2\ell + 1$ and let Q be the permutation matrix corresponding to the permutation $\pi(m) := 2m \pmod{n}$, $m = 0,1,\dots,n-1$. Then $QP_nQ^T = P_n^2$ and $Q(I + P_n + P_n^{\ell+1})Q^T = I + P_n + P_n^2$. Using the notation

(5.5.28) $$p_n(k_1,k_2,\dots,k_\ell) := \operatorname{per}(P_n^{k_1} + P_n^{k_2} + \dots + P_n^{k_\ell})$$

we have shown

$$p_{2\ell+1}(0,1,\ell+1) = p_{2\ell+1}(0,1,2) .$$

If k and ℓ are fixed then $p_n(0,k,\ell)$ grows exponentially:

THEOREM 5.5.5. *If* $0 < k < \ell$ *then*

(5.5.29) $$p_n(0,k,\ell) \geq 2^{[n/\ell]} .$$

Proof. The proof depends on (5.5.3). It is best illustrated by a figure only.

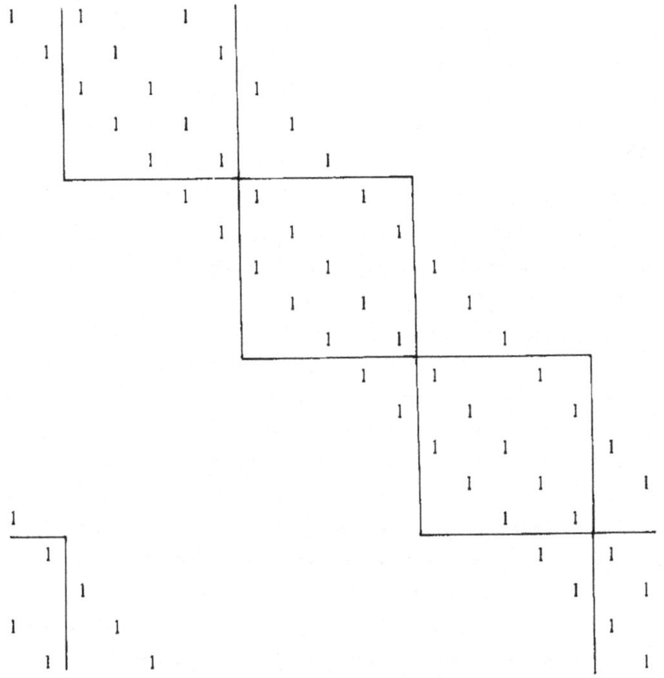

figure 22

In figure 22 we can replace a number of 1's by 0's transforming $I_{19} + P_{19}^2 + P_{19}^5$ into the direct sum of I_4 and three copies of $I_5 + P_5^3$.

The main purpose of this part of our research was an attempt to find something better than (5.5.26) for the class of circulants $C_n := \{I + P_n^k + P_n^\ell \mid 0 < k < \ell < n\}$. The partial results mentioned below would become a lot more valuable if we could prove the following conjecture:

CONJECTURE. *For every* ℓ *there is a number* ℓ_0 *such that* $p_n(0,1,\ell)$ *increases with* n *for* $n > \ell_0$. (In fact we conjecture $\ell_0 = 2\ell$.)

It is easily seen that in studying lower bounds for permanents of matrices in C_n we can take $k = 1$ (by a suitable permutation of rows and columns). If the conjecture is true then a suitable permutation of rows and columns can be used to show that it is no loss of generality to assume $\ell < n^{\frac{1}{2}}$ and then (5.5.29) shows that the minimal permanent in C_n increases at least as fast as $\exp(n^{\frac{1}{2}})$. One other method of attack is calculating the corresponding determinants where, if necessary, 1's in the matrix can be replaced by -1's. We then find lower bounds for the permanents. We use the following well known lemma.

LEMMA. *With* $\xi_k := \exp(2\pi ik/n)$ *we have*

(5.5.30)
$$\det\Big(\sum_{j=0}^{n-1} a_j P_n^j\Big) = \prod_{k=0}^{n-1} \Big(\sum_{j=0}^{n-1} a_j \xi_k^j\Big) .$$

Let $\alpha_1,\alpha_2,\dots,\alpha_\ell$ be the zeros of $z^\ell - z - 1$, i.e. $z^\ell - z - 1 = \prod_{j=1}^{\ell} (z - \alpha_j)$. Then we have by (5.5.30)

(5.5.31)
$$p_n(0,1,\ell) \ge |\det(- I - P_n + P_n^\ell)| = \prod_{k=0}^{n-1} |\xi_k^\ell - \xi_k - 1| =$$
$$= \prod_{k=0}^{n-1} \prod_{j=1}^{\ell} |\xi_k - \alpha_j| = \prod_{j=1}^{\ell} |\alpha_j^n - 1| .$$

Now take ℓ fixed. Since $z^\ell - z - 1$ has no zeros on the unit circle we find from (5.5.31)

(5.5.32)
$$\theta_\ell := \varliminf_{n\to\infty} \{p_n(0,1,\ell)\}^{1/n} \ge \prod_{j=1}^{\ell} \operatorname{Max}\{1,|\alpha_j|\} =: L_\ell .$$

From (5.5.29) we already know that $\theta_\ell \ge 2^{1/\ell}$ but we can even show that there is a constant $L > 1$ such that $L_\ell \ge L$ for all ℓ. (This is what we would expect from the Van der Waerden conjecture.) In $z^\ell - z - 1 = 0$ we substitute $z = \rho e^{i\theta}$. This yields $\rho^\ell \cos \ell\theta = \rho \cos\theta + 1$ and $\rho^\ell \sin \ell\theta = \rho \sin\theta$, and therefore $\rho^{2\ell} = \rho^2 + 1 + 2\rho \cos\theta$. This is the equation in polar coordinates of a curve K in the complex plane on which the zeros α_j all lie.

For any θ between $-\frac{\pi}{2}$ and $\frac{\pi}{2}$ the equation $x^{2\ell} - x^2 - 2x\cos\theta - 1 = 0$ has one positive zero and by substitution we easily see that this zero is between $1 + \frac{1}{4}\ell^{-1}$ and $1 + \ell^{-1}$. Considering the equation $(\sin \ell\theta)/(\sin\theta) = \rho^{1-\ell}$ along the part of K in the right half-plane we see that $z^{\ell} - z - 1$ has $2[\frac{\ell - 1}{4}]$ zeros in the right half-plane. Now from (5.5.32) we have

$$L_{\ell} > (1 + \tfrac{1}{4}\ell^{-1})^{2[\frac{\ell-1}{4}]} \to e^{1/8} \qquad (\ell \to \infty) .$$

This proves the assertion.
The investigations on permanents of circulants are being continued.

5.6. *Partial Latin squares.*

One of the consequences of the theorems of Hall and König is H. Theorem 5.1.5 which states that if a number of rows of a "possible" Latin square are given, then the square can be completed. We now look at other types of partial Latin squares which we define as follows:

DEFINITION. *Let* A *be an* n *by* n *matrix with entries* 0,1,2,...,n. A *is called a partial Latin square of order* n *if there is a Latin square* $B = [b_{ij}]$ *such that*

$$\forall_i \forall_j [a_{ij} \neq 0 \Rightarrow a_{ij} = b_{ij}] .$$

DEFINITION. *Let* A *be an* n *by* n *matrix with entries* 0,1,2,...,n. A *is called an incomplete Latin square if no positive integer occurs more than once in a row or column of* A.

An obvious question now is to find conditions under which an incomplete Latin square is a partial Latin square and also conditions for the contrary. We mention an open problem.

CONJECTURE. *An incomplete Latin square of order* n *with less than* n *nonzero entries is a partial Latin square.*

The following two examples show that there are incomplete Latin squares with n nonzero entries which are not partial Latin squares:

(i) $a_{ij} = n$ for $1 \le i = j \le \ell$, $a_{\ell+1,\ell+i} = i$ $(i = 1,2,\ldots,n-\ell)$;

(ii) $a_{i1} = i$ for $1 \le i \le \ell$, $a_{\ell+1,j} = \ell + j - 1$ for $j = 2,3,\ldots,n-\ell+1$.

In case (i) there is no place for the entry n in row $\ell + 1$. In case (ii) there is no nonzero entry possible for position $(\ell+1,1)$. It has been conjectured (D.A. Klarner, oral communication) that any incomplete Latin square which is not a partial Latin

square contains a configuration of type (i) or (ii) (after suitable permutations etc.). One has to interpret this conjecture rather liberally. E.g., if A is an m by m Latin square on the symbols 1,2,...,m and A is bordered by rows and columns of zeros to give it size n by n, where $n < 2m$, then obviously we have an incomplete Latin square which is not a partial Latin square. This example differs from (i) and (ii) above. However, no matter what we insert in positions $(m+1,j)$ $(j = 2,3,\dots)$, these positions together with the first column of A yield configuration (ii).
Concerning incomplete Latin squares in which the nonzero entries form a rectangle there is an interesting theorem of H.J. Ryser ([7]). We present a proof depending on the theorem of H. Ch.5.

THEOREM 5.6.1. *Let* A *be an incomplete Latin square of order* n *in which* $a_{ij} \neq 0$ *if and only if* $i \le r$ *and* $j \le s$. *Then* A *is a partial Latin square if and only if* $N(i) \ge r + s - n$ $(i = 1,2,\dots,n)$ *where* N(i) *is the number of elements of* A *equal to* i.

Proof.

(i) Let B be the r by n (0,1) matrix defined by

$(b_{ij} := 1) \iff$ (j does not occur in row i of A) , $(i = 1,\dots,r;\ j = 1,\dots,n)$.

Clearly every row of B has sum $n-s$. The j-th column of B has sum $r - N(j) \le n-s$. Therefore it is possible to add $n-r$ rows of nonnegative integers to B such that the resulting matrix B^* has all rowsums and columnsums equal to $n-s$. A consequence of H. Theorem 5.1.9 (see the proof given there) is that B^* is the sum of $n-s$ permutation matrices. Hence

$$B = L^{(s+1)} + \dots + L^{(n)}, \qquad (L^{(t)} = [\ell_{ij}^{(t)}]),$$

where $L^{(t)}$ is an r by n (0,1) matrix with one 1 in each row. In A we replace the 0 in position (i,j) $(i = 1,\dots,r;\ j = s+1,\dots,n)$ bij k if $\ell_{ik}^{(j)} = 1$. Thus A is changed into an incomplete Latin square of order n with r complete rows. By H. Theorem 5.1.5 this is a partial Latin square.

(ii) To prove the "only if" part we remark that in a Latin square of order n the first r rows contain exactly r i's and the last $n-s$ columns contain at most $n-s$ i's. Hence $N(i) \ge r - (n-s)$ is necessary.

5.7. *A matching problem.*

Consider the set $N := \{1,2,\dots,n\}$. We represent the subsets of N as vertices of a graph G_n (with 2^n vertices) and join two vertices by an edge iff one of the corresponding subsets is properly contained in the other. Let $\mathcal{A}_k := \{A \subset N \mid |A| = k\}$ and use the same notation for the corresponding subset of vertices of G_n. The sets $\mathcal{A}_k$ and $\mathcal{A}_{n-k}$ both contain $\binom{n}{k}$ vertices of G_n. A (complete) *matching* of these sets is a one-one correspondence between the vertices in $\mathcal{A}_k$ and $\mathcal{A}_{n-k}$ with the property that corresponding vertices are joined. Since every vertex in $\mathcal{A}_k$ is joined to $\binom{n-k}{k}$ vertices of $\mathcal{A}_{n-k}$ and vice versa, it follows from Hall's theorem (H. Theorem 5.1.1) that such a matching indeed exists. The question was raised to find a "natural" one-one correspondence which achieves such a matching. We shall now describe such a correspondence. Without loss of generality $k \le \frac{1}{2}n$. It is sufficient to find a one-one mapping of $\mathcal{A}_k$ onto $\mathcal{A}_k$ such that corresponding sets have empty intersections. We place the points $1,2,\dots,n$ on a circle. In the following we use addition mod n, i.e. $n+1 = 1$. Let $A = \{a_1,a_2,\dots,a_k\}$ where $1 \le a_1 < a_2 < \dots < a_k \le n$. Starting at a_1 we go around the circle (possibly more than once) and each time a point x not in A is passed we decide whether to include it in a set A^*. The rule is that if we have passed ℓ distinct points of A and have defined less than ℓ elements of A^*, then x is put in A^*. Otherwise we do not, although it may be necessary to include x in A^* the next time around. The process is illustrated by the example $n = 14$, $A := \{1,2,4,9,14\}$ where we find $3 \in A^*$, $5 \in A^*$, $6 \in A^*$, $10 \in A^*$, $7 \in A^*$. Clearly the process terminates when k elements of A^* have been defined. Then $|A| = |A^*|$ and $A \cap A^* = \emptyset$. Define $A^{**} = N \setminus (A \cup A^*)$. If x, y are two consecutive points of A^{**} on the circle then for $x < \xi < y$ we have

$$|[x+1,\xi] \cap A| \ge |[x+1,\xi] \cap A^*| ,$$

equality holding if $\xi = y-1$. This shows us two things:

(i) If we start the process described above at some other element of A than a_1 we find the same set A^* (because elements of A^{**} clearly remain in A^{**}).

(ii) If we start with A^*, reverse the order in which we go around the circle and then apply the same rule as before, we find A.

From (ii) it follows that $\varphi(A) := A^*$ defines a one-one mapping of $\mathcal{A}_k$ onto $\mathcal{A}_k$. For every $A \in \mathcal{A}_k$ we have $\varphi(A) \cap A = \emptyset$, i.e. $A \subset N \setminus \varphi(A)$. This is the "natural" mapping.

Let us consider the graph G_n once more. If a sequence of points P_i $(i = k,k+1,\dots,n-k)$ has the properties $P_i \in \mathcal{A}_i$ $(i = k,k+1,\dots,n-k)$ and P_iP_{i+1} is an edge in G_n $(i = k,\dots,n-k-1)$ then we call this sequence a *symmetric chain* in G_n. A natural algorithm for splitting the graph G_n into symmetric chains was described by N.G. de Bruijn, C. van Ebbenhorst Tengbergen and D. Kruyswijk [18]. It starts with one chain for $n = 1$ and then proceeds by induction. Suppose for some n the graph G_n has been split up into symmetric chains and let $P_k,P_{k+1},\dots,P_{n-k}$ be such a chain; (we remind the reader that P_i stands for a point in G_n and also for a subset of $\{1,2,\dots,n\}$ of cardinality i). Then $P_{k+1},P_{k+2},\dots,P_{n-k}$ and $P_k,P_k \cup \{n+1\},P_{k+1} \cup \{n+1\},\dots,P_{n-k} \cup \{n+1\}$ are symmetric chains in G_{n+1}. The procedure is illustrated in figure 23.

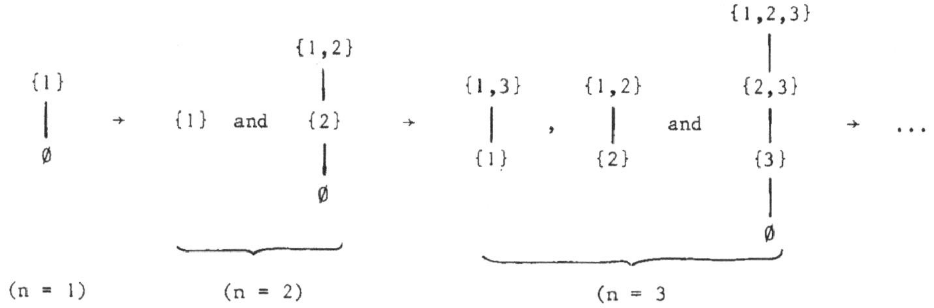

figure 23

Once G_n has been split up into symmetric chains we have a matching for $\mathcal{A}_k$ and $\mathcal{A}_{n-k}$, the correspondence being belonging to the same symmetric chain. The reader is invited to check that the matching derived from [18] is the same as the one we described directly above. If $k \le \frac{1}{2}(n-1)$ then $|\mathcal{A}_k| \le |\mathcal{A}_{k+1}|$. A matching from $\mathcal{A}_k$ to $\mathcal{A}_{k+1}$ defined by ordering both sets lexicographically and then mapping sequentially to the first "possible" point was described by M. Lewin [19]. Again this matching coincides with the one described in figure 23.

5.8. *Theorems of Birkhoff and Caratheodory.*

Problem 4 on H.p.53 suggests the use of a theorem of Caratheodory which says that if $\underline{a}_1, \underline{a}_2, \ldots, \underline{a}_k$ $(k \ge n+1)$ are points in $\mathbb{R}^n$, then every point $\underline{a}$ in the convex hull of these points can be written as $\underline{a} = \sum_{i=1}^{k} \alpha_i \underline{a}_i$ with at most $n+1$ nonzero coefficients and $\sum_{i=1}^{k} \alpha_i = 1$. However, this would yield the bound $n^2 - 2n + 3$ instead of $n^2 - 2n + 2$. Therefore we first prove a useful modification of Caratheodory's theorem.

DEFINITION. *A set* $C \subset \mathbb{R}^n$ *is called a convex cone if* $\lambda\underline{x} + \mu\underline{y} \in C$ *whenever* $\underline{x}, \underline{y} \in C$; $\lambda, \mu \ge 0$. *The convex cone generated by a set* S *is defined to be the intersection of the set of convex cones containing* S *(notation:* cc(S)*).*

Obviously, cc(S) is a convex cone. It is easily seen that cc(S) is the set of all linear combinations with nonnegative coefficients of elements of S. Caratheodory's theorem gives an upper bound for the number of elements of S which are needed.

THEOREM 5.8.1. *If* $\underline{x} \in cc(S)$, *then there exist positive numbers* $\lambda_1, \lambda_2, \ldots, \lambda_m$ *and a sequence of independent vectors* $\underline{x}_1, \underline{x}_2, \ldots, \underline{x}_m$ *in* S *such that* $\underline{x} = \sum_{k=1}^{m} \lambda_k \underline{x}_k$.

Proof. We have already seen that $\underline{x}$ can be written as $\underline{x} = \sum_{k=1}^{m} \lambda_k \underline{x}_k$ with $\lambda_k \ge 0$, $\underline{x}_k \in S$ $(k = 1,2,\ldots,m)$. Of all possible representations of this form we choose one which minimizes m. Then obviously $\lambda_k > 0$ $(k = 1,2,\dots,m)$. Suppose that $\underline{x}_1, \underline{x}_2, \dots, \underline{x}_m$ are dependent, say $\sum_{k=1}^{m} \alpha_k \underline{x}_k = \underline{0}$, where at least one α_k is positive. Then we define

$$\rho := \min\{\lambda_k/\alpha_k \mid \alpha_k > 0\} .$$

It follows that $\lambda_k - \rho\alpha_k \geq 0$ for $k = 1,2,\dots,m$, and that $\lambda_k - \rho\alpha_k = 0$ for at least one value of k. Furthermore we have $\underline{x} = \sum_{k=1}^{m} (\lambda_k - \rho\alpha_k)\underline{x}_k$ in which we can delete the term with $\lambda_k - \rho\alpha_k = 0$ contradicting the definition of m.

COROLLARY. *If* $S \subset \mathbb{R}^n$ *then* $cc(S) = \{\sum_{k=1}^{n} \lambda_k \underline{x}_k \mid \lambda_k \geq 0,\ \underline{x}_k \in S\ (k = 1,2,\dots,n)\}$.

Now Birkhoff's theorem (H. Theorem 5.1.9) states that the set of matrices with non-negative entries for which every row and column has the same sum is the convex cone generated by the permutation matrices. Since the set of matrices is a vector space of dimension $n^2 - 2n + 2$ the result follows from Theorem 5.8.1.

References.

[1] M. Hall, Distinct Representatives of Subsets, Bull. Am. Math. Soc. 54 (1948), 922-926.

[2] R. Rado, On the Number of Systems of Distinct Representatives of Sets, J. London Math. Soc. 42 (1967), 107-109.

[3] P.A. Ostrand, Systems of Distinct Representatives, II, J. of Math. Analysis and Applic. 32 (1970), 14.

[4] P. Hall, On Representatives of Subsets, J. London Math. Soc. 10 (1935), 26-30.

[5] P.R. Halmos and H.E. Vaughan, The Marriage Problem, Am. J. of Math. 72 (1950), 214-215.

[6] H.J. Ryser, Permanents and Systems of Distinct Representatives, Chapter 4 of Combinatorial Mathematics and its Applications, University of North Carolina Press (1969).

[7] R. Rado, A Theorem on Independence Relations, Quart. J. Math. (Oxford) 13 (1942), 83-89.

[8] M.L.J. Hautus, Stabilization, Controllability and Observability of Linear Autonomous Systems, Nederl. Akad. Wetensch. Proc. Ser. A 73 (1970), 448-455.

[9] M. Hall, An Algorithm for Distinct Representatives, Am. Math. Monthly 63 (1956), 616.617.

[10] V. Chvátal, Distinct Representatives for a Collection of Finite Sets, Am. Math. Monthly 79 (1972), 775.

[11] H. Minc, An Inequality for Permanents of (0,1) Matrices, J. Comb. Theory 2 (1967), 321-326.

[12] H. Minc, Upper Bounds for Permanents of (0,1) Matrices, Bull. Am. Math. Soc. 69 (1963), 789-791.

[13] W.B. Jurkat and H.J. Ryser, Matrix Factorizations of Determinants and Permanents, J. Algebra 3 (1966), 1-27.

[14] D.J. Hartfiel and J.W. Crosby, A Lower Bound for the Permanent on $U_n(k,k)$, J. Comb. Theory (A) 12 (1972), 283-288.

[15] D.J. Hartfiel, A Simplified Form for Nearly Reducible and Nearly Decomposable Matrices, Proc. Am. Math. Soc. 24 (1970), 388-393.

[16] A. Nijenhuis and H.S. Wilf, On a Conjecture of Ryser and Minc, Nederl. Akad. Wetensch. Proc. Ser. A 73 (1970), 151-157.

[17] H.J. Ryser, A Combinatorial Theorem with an Application to Latin Rectangles, Proc. Am. Math. Soc. 2 (1951), 550-552.

[18] N.G. de Bruijn, C. van Ebbenhorst Tengbergen and D. Kruyswijk, On the Set of Divisors of a Number, Nieuw Archief v. Wisk. (2) 23 (1951), 191-193.

[19] M. Lewin, Choice Mappings of Certain Classes of Finite Sets, Math. Z. 124 (1972), 23-36.

VI. RAMSEY's THEOREM

6.1. *Introduction; elementary theorems.*

We first state Ramsey's theorem in the general form.

THEOREM 6.1.1. *Let* $r \geq 1$, $q_i \geq r$ $(i = 1,2,\dots,s)$. *Let* S *be a set of* n *elements and suppose that the family* T *of all subsets of* S *containing exactly* r *elements is divided into* s *mutually exclusive families* $T_1,T_2,\dots,T_s$. *Then there exists a minimal positive integer* $N(q_1,q_2,\dots,q_s;r)$ *such that, if* $n \geq N(q_1,q_2,\dots,q_s;r)$, *then there is an* i $(1 \leq i \leq s)$ *such that* S *has a* q_i*-subset all of whose* r*-subsets are in* T_i.

Note that H. Theorem 6.1.1 is the case $s = 2$. In this chapter we shall only consider the case $r = 2$. The theorem then states that if the edges of the complete graph K_n on n vertices are colored with s colors and $n \geq N(q_1,q_2,\dots,q_s;2)$, then there is an i such that K_n has a complete subgraph on q_i points with all edges in the i-th color. We shall call this a *monochromatic* complete subgraph K_{q_i} of color i. If $s = 2$ we shall always take red and blue for the colors (in this order). The points of K_n will be denoted by $P_1,P_2,\dots,P_n$. We use the following notation and terminology:

(a) If K_n is colored red and blue then r_i denotes the number of red edges with P_i as endpoint.

(b) We say K_n is (p,q)-*colored* if it contains no red K_p and no blue K_q.

(c) The maximal n for which a (p,q)-coloring of K_n is possible is $R(p,q) = N(p,q;2) - 1$.

(d) $r(n;p,q)$ is the maximal number of red edges in a K_n which is (p,q)-colored; if $n > R(p,q)$ then $r(n;p,q) := 0$. For the color blue $b(n;p,q)$ is defined analogously. Hence $b(n;p,q) = r(n;q,p)$.

The following two equations are obvious:

$$(6.1.1) \qquad N(p,q;1) = p + q - 1$$

$$(6.1.2) \qquad N(p,2;2) = p \ .$$

The following lemma shows the simplest argument occurring in proofs of inequalities for Ramsey numbers.

LEMMA 6.1.1. *If* K_n *is* (p,q)*-colored then we have*

$$(6.1.3) \qquad r_i \leq R(p-1,q) \qquad (i = 1,2,\dots,n) \ .$$

Proof. By definition a K_{r_i} with $r_i > R(p-1,q)$ contains a monochromatic red K_{p-1} or a blue K_q. If the points of this K_{r_i} are joined by red edges to P_i then we have a K_{r_i+1}

in K_n which contains a red K_p or a blue K_q.

Corollary.

(6.1.4) $$r(n;p,q) \le \tfrac{1}{2}n\, R(p-1,q) \ ,$$

(6.1.5) $$b(n;p,q) \le \tfrac{1}{2}n\, R(p,q-1) \ .$$

Now assume K_n is (p,q)-colored. Then counting the edges gives us

$$\tfrac{1}{2}n(n-1) \le r(n;p,q) + b(n;p,q) \le \tfrac{1}{2}n\, R(p-1,q) + \tfrac{1}{2}n\, R(p,q-1) \ ,$$

i.e.

(6.1.6) $$n \le R(p-1,q) + R(p,q-1) + 1 \ .$$

Equality in (6.1.6) is possible only if both the number of red edges and the number of blue edges are maximal. Suppose $R(p-1,q)$ and $R(p,q-1)$ are both odd. Then equality in (6.1.4) and (6.1.5) is possible only if n is even and then (6.1.6) cannot be an equality. We have proved:

THEOREM 6.1.2. *The Ramsey numbers* $N(p,q;2)$ *satisfy*

(6.1.7) $$N(p,q;2) \le N(p-1,q;2) + N(p,q-1;2)$$

with strict inequality if $N(p-1,q;2)$ *and* $N(p,q-1;2)$ *are both even.* (See [1].)

Corollary.

(6.1.8) $$N(p,q;2) \le \binom{p+q-2}{p-1} \ .$$

Proof. This follows from (6.1.2), (6.1.7) and the fact that the binomial coefficients satisfy (6.1.7) with equality.

Problem 1 on H.p.57 is the simplest example of a theorem by A.W. Goodman (cf. [2], [3]) which we now prove.

THEOREM 6.1.3. *If the edges of* K_n *are colored red or blue and* r_i $(i = 1,2,\ldots,n)$ *denotes the number of red edges with* P_i *as endpoint and if* Δ *denotes the number of monochromatic triangles (red or blue) in* K_n*, then*

(6.1.9) $$\Delta = \binom{n}{3} - \tfrac{1}{2}\sum_{i=1}^{n} r_i(n-1-r_i) \ .$$

Proof. Every triangle in K_n which is not monochromatic has exactly two vertices where a red and a blue edge meet. In P_i two such edges can be chosen in $r_i(n-1-r_i)$ ways. Hence $\sum_{i=1}^{n} r_i(n-1-r_i)$ counts the non-monochromatic triangles twice.

Corollary.

$$\Delta \geq \binom{n}{3} - [\tfrac{n}{2}\,[(\tfrac{n-1}{2})^2]] \,. \qquad (6.1.10)$$

Proof. Δ in (6.1.9) is minimized if $r_i = n - r_i - 1$ for all i (n odd), respectively $r_i = \frac{n}{2}$ or $\frac{n}{2} - 1$ for all i (n even). Since Δ is an integer we find (6.1.10). It is easy to find examples which show that (6.1.10) cannot be improved (cf. [3]).

The following lemma gives an upper bound for the number Δ (cf. [4]).

LEMMA 6.1.2. *Let* K_n *be* (p,q)*-colored. Let there be* α_j *vertices with* $r_i = j$ $(j = 0,1,\dots,n-1)$. *Then*

$$\Delta \leq \frac{1}{3} \sum_{j=0}^{n-1} \alpha_j \{r(j;p-1,q) + b(n-1-j;p,q-1)\} \,. \qquad (6.1.11)$$

Proof. Let $r_i = j$. Then P_i is joined by red edges to a K_j which is (p-1,q)-colored since K_n is (p,q)-colored. Therefore P_i is vertex of at most r(j;p-1,q) monochromatic red triangles. The same reasoning applies to the blue edges from P_i.

We remark that by (6.1.3) we have $\alpha_j = 0$ unless $n - 1 - R(p,q-1) \leq j \leq R(p-1,q)$. The two inequalities for Δ enable us to prove a theorem of the same type as Theorem 6.1.2.

THEOREM 6.1.4. *Let* K_n *be* (p,q)*-colored and let* $R_1 := R(p-2,q)$ *and* $R_2 := R(p,q-2)$. *Then*

$$n \leq R_1 + R_2 + 3 + 2\{\tfrac{1}{3}(R_1^2 + R_1R_2 + R_2^2) + R_1 + R_2 + 1\}^{\frac{1}{2}} \,. \qquad (6.1.12)$$

Proof. By combining (6.1.9), (6.1.11) and (6.1.4) and (6.1.5) we find

$$\binom{n}{3} \leq \frac{1}{2} \sum_{i=1}^{n} r_i(n-1-r_i) + \frac{1}{6} \sum_{j=0}^{n-1} \alpha_j\{jR_1 + (n-1-j)R_2\} \,.$$

Since $\sum_{j=0}^{n-1} \alpha_j = n$ and $\sum_{j=0}^{n-1} j\alpha_j = \sum_{i=1}^{r} r_i$, the right hand side is a quadratic form in the variables r_i. If we maximize this and solve for n we find (6.1.12).

Corollary (cf. [4]).

$$N(p,p;2) \leq 4N(p,p-2;2) + 2 \,. \qquad (6.1.13)$$

Proof. In (6.1.12) take $R_1 = R_2 = N(p,p-2;2) + 1$.

Remark. If in (6.1.12) we replace R_1 by $N_1 - 1$ and R_2 by $N_2 - 1$, where $N_1 := N(p-2,q;2)$, $N_2 := N(p,q-2;2)$, we obtain

$$N(p,q;2) \le N_1 + N_2 + 2 + 2\left(\frac{N_1^2 + N_1N_2 + N_2^2}{3}\right)^{\frac{1}{2}}$$

as a generalization of (6.1.13).

6.2. *Some values of* N(p,q;2).

If p or q is 1 or 2 then we know N(p,q;2) by (6.1.1) and (6.1.2). By (6.1.7) we have $N(3,3;2) \le 6$. If we color the edges of a pentagon red and the diagonals blue we have a (3,3)-colored K_5. This shows that $N(3,3;2) = 6$. Again by (6.1.7) we have $N(3,4;2) \le 9$. An example of a (3,4)-colored K_8 is given in H.p.57 Problem 7. Hence $N(3,4;2) = 9$. By (6.1.7) $N(3,5;2) \le 14$. Consider K_{13}. Let P_iP_j be colored red if and only if $i - j \equiv \pm 1$ or $\pm 5 \pmod{13}$. By inspection it is easily checked that this is a (3,5)-coloring. Hence $N(3,5;2) = 14$. By H.p.57 Problem 8 we have $N(4,4;2) = 18$. There are only two more Ramsey numbers which are known, namely $N(3,6;2) = 18$, $N(3,7;2) = 23$ (cf. [5], [6]). These known values and (6.1.7), (6.1.12), (6.1.13), etc., lead to inequalities $N(4,5;2) \le 30$, $N(5,5;2) \le 58$, etc. Below we list the bounds known at the moment for some of the smaller Ramsey numbers (cf. [7]). These are obtained by a closer inspection of the possible values of r_i etc. than in the crude argument of Theorem 6.1.4.

Table of N(p,q;2)

p = 6	6	18			102/178				
5	5	14		38/55	38/94				
4	4	9	18	25/28	34/45				
3	3	6	9	14	18	23	27/30	36/37	
2	2	3	4	5	6	7	8	9	10
q →	2	3	4	5	6	7	8	9	10

Even for small values of p and q, e.g. p = q = 5, the bounds are very far apart and the problem of determining new Ramsey numbers does not look attractive at the moment.

6.3. *The numbers* N(p,p;2).

The inequality (6.1.8) shows that $N(p,p;2) \le \binom{2p-2}{p-1}$ and hence

(6.3.1) $$N(p,p;2) \le 2^{2p},$$

in fact even $N(p,p;2) = o(2^{2p})$ $(p \to \infty)$. We now give a lower bound due to Erdős ([8]). There are $2^{\binom{n}{2}}$ possible ways of coloring K_n. Given a specific K_k in K_n, there are

$2^{1+\binom{n}{2}-\binom{k}{2}}$ ways of coloring K_n such that the given K_k is monochromatic. There are $\binom{n}{k}$ ways of choosing a K_k in K_n. It follows that a (k,k)-coloring of K_n exists if $2^{\binom{n}{2}} > \binom{n}{k}2^{1+\binom{n}{2}-\binom{k}{2}}$. Since $\binom{n}{k} < n^k/k!$ this is true if $n < (k!\,2^{\binom{k}{2}-1})^{1/k}$. This is so if $k \geq 3$, $n \leq 2^{k/2}$. Therefore we have proved:

THEOREM 6.3.1. *For* $k \geq 3$ *we have*

$$N(k,k;2) > 2^{k/2} . \tag{6.3.2}$$

From (6.3.1) and (6.3.2) we see, that

$$2^{\frac{1}{2}} \leq \{N(p,p;2)\}^{1/p} \leq 2 \qquad (p \geq 2) . \tag{6.3.3}$$

It is tempting to conjecture that $\{N(p,p;2)\}^{1/p}$ tends to a limit if $p \to \infty$. A very bold conjecture from the table in Section 6.2 is $N(p,p;2) = 2.3^{p-2}$ (H.J.L. Kamps). Recently, an improvement of (6.3.1) was found by J. Yackel ([9]), namely

$$N(p,p;2) = O\left(\frac{\log\log p}{\log p}\binom{2p-2}{p-1}\right) \qquad (p \to \infty) . \tag{6.3.4}$$

Of course this does not change the right hand side of (6.3.3).

6.4. *Inequalities for* N(p,3;2).

The results discussed above suggest that the inequality (6.1.8) with $q = 3$, i.e. $N(p,3;2) \leq \binom{p+1}{2}$, is probably not very good. The methods we have described were all rather elementary and therefore it is not surprising that the best presently known estimates for N(p,3;2) were derived by much more complicated methods. In 1960 Erdős ([10]) proved the inequality

$$N(p,3;2) > cp^2(\log p)^{-2} , \tag{6.4.1}$$

(c a constant). At that time he thought that maybe even $N(p,3;2) > c_1 p^2$ for some c_1, which would mean that (6.1.8) gives the correct order of growth. However, Graver and Yackel ([6]) proved in 1968 that

$$N(p,3;2) = O\left(p^2 \frac{\log\log p}{\log p}\right) \qquad (p \to \infty) . \tag{6.4.2}$$

Indeed, this shows that (6.1.8) is not a good estimate for large p. The result (6.4.2) was generalized in [6] to

$$N(p,q;2) \leq Cp^{q-1} \frac{\log\log p}{\log p} \qquad (q \geq 3) . \tag{6.4.3}$$

6.5. *Turan's theorem.*

Ramsey's theorem is sometimes formulated in the following way: If $n \geq N(p,q;2)$ then a graph on n vertices contains a complete graph on p vertices as a subgraph or a set of q vertices with no edges joining any pair of these q vertices. Clearly this formulation is obtained by coloring K_n and omitting the blue edges. A natural question to ask is whether some condition on the edges of a graph on n vertices guarantees that the graph contains a complete graph on p vertices. A theorem of this type was proved by P. Turan ([11]) and it seems worthwhile to mention it at this point.

THEOREM 6.5.1. *Consider all graphs on* n *vertices which do not contain a* K_k *as a subgraph. Let* $M(n,k)$ *be the maximal number of edges occurring in these graphs. If* r *is defined by* $n = t(k-1) + r$, $1 \leq r \leq k-1$, *then*

$$M(n,k) = \frac{k-2}{2(k-1)}(n^2 - r^2) + \binom{r}{2} . \tag{6.5.1}$$

Proof. (6.5.1) is obvious for $t = 0$. We proceed by induction on t. Clearly a graph which contains no K_k and which has the maximal number of edges must contain a K_{k-1}. Each of the remaining $n-k+1$ points is joined by an edge to at most $k-2$ of the points of this K_{k-1}. Furthermore, the remaining $n-k+1$ are mutually joined by at most $M(n-k+1,k)$ edges. Hence

$$M(n,k) \leq \binom{k-1}{2} + (k-2)(n-k+1) + M(n-k+1,k) . \tag{6.5.2}$$

Since the right hand side of (6.5.1) satisfies (6.5.2) with equality, we see that $M(n,k)$ is at most equal to the right hand side of (6.5.1). Now let $n = t(k-1) + r$. Divide a set of n points into r classes $C_1,\dots,C_r$ of $t+1$ points each and $k-1-r$ classes $C_{r+1},\dots,C_{k-1}$ of t points each. Points in the same class are not joined by an edge and points in different classes are joined by an edge. This graph on n points obviously contains no K_k as a subgraph and the number of edges is equal to the right hand side of (6.5.1). This proves the theorem.

Despite the analogy this theorem does not give us any information on the Ramsey numbers.

6.6. *Infinite graphs.*

It is possible to give a generalization of Ramsey's theorem to infinite graphs. The statement is that if the edges of the complete graph on the points $P_1,P_2,\dots$ (a countable set) are colored red and blue then there is an infinite subset with all joining edges red or all edges blue. For any set A let $P_2(A)$ denote the set of 2-subsets of A (edges). If $G \subset P_2(S)$ and $A \subset S$ then we write $G|A$ for $G \cap P_2(A)$ (restriction). Then we have:

THEOREM 6.6.1. *If* S *is a countable set and* $G \subset P_2(S)$ *then*

$$\exists_{A \subset S} [|A| = \infty \wedge G|A = P_2(A)] \vee \exists_{B \subset S} [|B| = \infty \wedge G|B = \emptyset] .$$

Proof. Let $S = \{P_n \mid n \in \mathbb{N}\}$. G is a graph on these points. The *valency* of P_n is the cardinality of $\{P_m \mid \{P_n, P_m\} \in G\}$.

(i) Suppose there is an infinite number of points P_n with finite valency, say $P_{i_1}, P_{i_2}, \ldots$. We construct a subsequence as follows. $Q_1 := P_{i_1}$, Q_2 is the first point in the sequence which is not joined to P_{i_1}. Then Q_3 is the first point following Q_2 which is not joined to Q_1 or Q_2, etc. Clearly we thus find an infinite subset of S such that no two points are joined by an edge of G.

(ii) We may now assume that for every infinite subset $A \subset S$ the graph $G|A$ has only a finite number of points of finite valency (within $G|A$). Let S_1 be the subset of those points of S which have infinite valency. Take $Q_1 \in S_1$ and let S_1' be the (infinite) subset of S_1 consisting of the points joined to Q_1. Let S_2 be the subset of S_1' consisting of the points of S_1' which have infinite valency in the graph $G|S_1'$. Now choose $Q_2 \in S_2$ and proceed in this way. We thus find an infinite complete subgraph of G.

Remark. The theorem is also true for more than two colors. This is easily proved by induction.

We give two interesting applications of this theorem.

THEOREM 6.6.2. *Let* $(a_n)_{n \in \mathbb{N}}$ *be a sequence of real numbers. Then* $(a_n)_{n \in \mathbb{N}}$ *has a strictly monotone subsequence or a constant subsequence.*

Proof. Take the terms of the graph as the vertices of an infinite complete graph. Color the edge $\{a_i, a_j\}$ red if $j > i$ and $a_j > a_i$, blue if $j > i$ and $a_j < a_i$, green if $a_i = a_j$. Since by Theorem 6.6.1 (remark) there is a monochromatic infinite subgraph the theorem is proved.

THEOREM 6.6.3. *Let* $(f_n)_{n \in \mathbb{N}}$ *be a uniformly bounded sequence of continuous real functions on* $[0,1]$. *Suppose there is an integer* $k \geq 0$ *such that for all* i, j $(i \neq j)$ *the equation* $f_i(x) = f_j(x)$ *has at most* k *solutions. Then there exists a subsequence which is convergent on* $[0,1]$.

Proof. The proof is by induction on k. For $k = 0$ the theorem is a consequence of Theorem 6.6.2. If a subsequence of $(f_n(0))_{n \in \mathbb{N}}$ is monotonic (say increasing) then this same subsequence is increasing in every point x and convergent because of boundedness. Suppose the theorem is true for $k-1$. Take the f_i as vertices of an infinite complete graph. The edge $\{f_i, f_j\}$ is red if the equation $f_i(x) = f_j(x)$ has exactly k solutions in the interval $[0,\frac{1}{2}]$. Otherwise the edge is blue. By Theorem 6.6.1 this graph has an infinite complete monochromatic subgraph. If this subgraph is red then we apply the theorem for k = 0 on $(\frac{1}{2},1]$. Otherwise we apply the induction hypothesis on $[0,\frac{1}{2}]$. We remark that if a sequence of functions is bounded and convergent on a set with the exception of one point then it has a subsequence convergent on the whole set. Hence in both cases we have found a subsequence convergent on half of [0,1]. Now divide the remaining half in two equal parts and proceed by induction. By the usual diagonal-procedure we thus find a sequence of functions which is convergent on [0,1], possibly with the exception of one point. We have already remarked that this sequence has a subsequence which is also convergent on the remaining point.

This elegant proof is not new but we do not know where it originates.

References.

[1] R.E. Greenwood and A.M. Gleason, Combinatorial Relations and Chromatic Graphs, Can. J. Math. 7 (1955), 1-7.

[2] A.W. Goodman, On Sets of Acquaintances and Strangers at any Party, Am. Math. Monthly 66 (1959), 778-783.

[3] A.J. Schwenck, Acquaintance Graph Party Problem, Am. Math. Monthly 79 (1972), 1113-1117.

[4] K. Walker, Dichromatic Graphs and Ramsey Numbers, J. Comb. Theory 5 (1968), 238-243.

[5] G. Kéry, Ramsey egy Gráfelméléti Tététeröl, Mat. Lapok. 15 (1964), 204-224.

[6] J.E. Graver and J. Yackel, Some Graph Theoretic Results Associated with Ramsey's Theorem, J. Comb. Theory 4 (1968), 125-175.

[7] K. Walker, An Upper Bound for the Ramsey Number M(5,4), J. Comb. Theory 11 (1971), 1-10.

[8] P. Erdös, Some Remarks on the Theory of Graphs, Bull. A.M.S. 53 (1947), 292-294.

[9] J. Yackel, Inequalities and Asymptotic Bounds for Ramsey Numbers, J. Comb. Theory (B), 13 (1972), 56-68.

[10] P. Erdös, Graph Theory and Probability II, Can. J. Math. 13 (1961), 346-352.

[11] P. Turan, Eine Extremalaufgabe aus der Graphentheorie, Math. Fiz. Lapok 48 (1941), 436-452.

VII. SOME EXTREMAL PROBLEMS

7.1. *The assignment problem.*

We remind the reader that an optimal assignment for the matrix $A := [a_{ij}]$ $(i = 1,\dots,n;$ $j = 1,\dots,n)$ is a permutation π which maximizes $\sum_{i=1}^{n} a_{i\pi(i)}$.

In connection with H.p.65 Problem 1 the question was raised whether in every optimal assignment for the matrix A there is at least one row such that a maximal element of this row occurs as one of the scores. We shall show the answer to be affirmative. The matrix $\begin{pmatrix} 0 & 0 & 1 \\ 0 & 0 & 1 \\ 0 & 0 & 1 \end{pmatrix}$ shows that we cannot make the same statement for more than one row.

THEOREM 7.1.1. *Let* $A = [a_{ij}]$ *be an* $n \times n$ *matrix of real numbers. Let the permutation* π *be such that the sum* $\sum_{i=1}^{n} a_{i\pi(i)}$ *is maximal. Then there is an* i $(1 \le i \le n)$ *such that*

$$\max\{a_{ij} \mid 1 \le j \le n\} = a_{i\pi(i)} .$$

Proof. Let u_i $(i = 1,\dots,n)$, v_j $(j = 1,\dots,n)$ be such that $u_i + v_j \ge a_{ij}$ for all i and j and that $\sum_{i=1}^{n} u_i + \sum_{j=1}^{n} v_j$ is minimal. We know by H. Theorem 7.1.1 that $a_{i\pi(i)} = u_i + v_{\pi(i)}$ for $i = 1,2,\dots,n$. Now let $\max\{v_j \mid j = 1,\dots,n\} =: v_k$. Choose i such that $\pi(i) = k$. Then

$$a_{i\pi(i)} = u_i + v_k = \max\{u_i + v_j \mid j = 1,2,\dots,n\}$$
$$\ge \max\{a_{ij} \mid j = 1,2,\dots,n\} \ge a_{i\pi(i)} .$$

7.2. *Dilworth's theorem.*

An elegant short proof of Dilworth's theorem for finite partially ordered sets P was given by H. Tverberg ([1]). We present the proof here. A subset of P whose elements are pairwise incomparable will be called an *antichain*.

THEOREM 7.2.1. *Given a partially ordered finite set* P. *The minimal number of disjoint chains which together contain all elements of* P *is equal to the maximal number of elements in an antichain in* P.

Proof. As in the proof of H. Theorem 7.2.1 we only have to show that if every antichain has at most k elements then P is the union of at most k disjoint chains. The proof is by induction on the number of elements of P. If $|P| = 0$ the theorem is trivial. Now let C be a maximal chain in P. If $P \setminus C$ has no antichain of k elements then we are finished. Assume $P \setminus C$ has an antichain of k elements, say $\{a_1,a_2,\dots,a_k\}$. Now define

$$S^- := \{x \in P \mid \exists_i \,[x \le a_i]\} ,$$

and S^+ analogously. Since C is a maximal chain, the largest element of C is not in S^- and hence by the induction hypothesis the theorem holds for S^-. Therefore S^- is the union of k disjoint chains, say $S_1^-, S_2^-, \ldots, S_k^-$, where $a_i \in S_i^-$. Suppose $x \in S_i^-$ and $x > a_i$. Since there is a j with $x \le a_j$, this would imply $a_i < a_j$, a contradiction. Hence a_i is the maximal element of the chain S_i^- $(i = 1,2,\ldots,k)$. In the same way S^+ is the union of k disjoint chains with the a_i as minimal elements. Combining the chains the theorem follows.

A dual of Dilworth's theorem was given recently by L. Mirsky ([2]). It is much easier to prove than Theorem 7.2.1.

THEOREM 7.2.2. *Let* P *be a partially ordered set. If* P *possesses no chain of* $m+1$ *elements, then* P *is the union of* m *antichains.*

Proof. For $m = 1$ the theorem is trivial. Let $m \ge 2$ and assume the theorem is true for $m-1$. Let P be a partially ordered set which has no chain of $m+1$ elements. Let M be the set of maximal elements of P. M is an antichain. Suppose $x_1 < x_2 < \ldots < x_m$ were a chain in $P \setminus M$. Then this would also be a maximal chain in P and hence $x_m \in M$, a contradiction. Hence $P \setminus M$ has no chain of m elements. By the induction hypothesis $P \setminus M$ is the union of $m-1$ antichains. This proves the theorem.

It is interesting to compare H.p.65 Problem 3, which is a consequence of Dilworth's theorem and also of the dual, with Ramsey's theorem. For example, consider a partially ordered set of 5 elements and join two elements by a red edge if they are comparable, by a blue edge otherwise. The result of the problem is that there is a monochromatic triangle, whereas $N(3,3;2) = 6$. Because of the transitivity not every coloring corresponds to a partial ordering!

We mention a problem connected to Section 5.7.

PROBLEM. Let $n = p_1^{\alpha_1} p_2^{\alpha_2} \ldots p_k^{\alpha_k}$. Wat is the maximal number of divisors of n such that no one of them divides any of the others?

Solution. The divisors of n form a partially ordered set as described in H. § 2.2, Case 2. A description is obtained by considering n as a "set" of $\alpha_1 + \alpha_2 + \ldots + \alpha_k$ prime factors and the divisors as subsets ordered by inclusion. We described in Section 5.7 that the corresponding graph is the union of symmetric chains. Here the word chain corresponds to the way we have used it in this chapter. This decomposition into chains shows that a maximal antichain is obtained by considering those divisors $d = p_1^{\beta_1} \ldots p_k^{\beta_k}$ for which $\beta_1 + \beta_2 + \ldots + \beta_k = \frac{\alpha_1 + \alpha_2 + \ldots + \alpha_k}{2}$ (cf. Chapter 5, reference [18]). We give an example in figures 24, 25 (see figure 23). Take

$n = 60 = 2^2.3.5.$

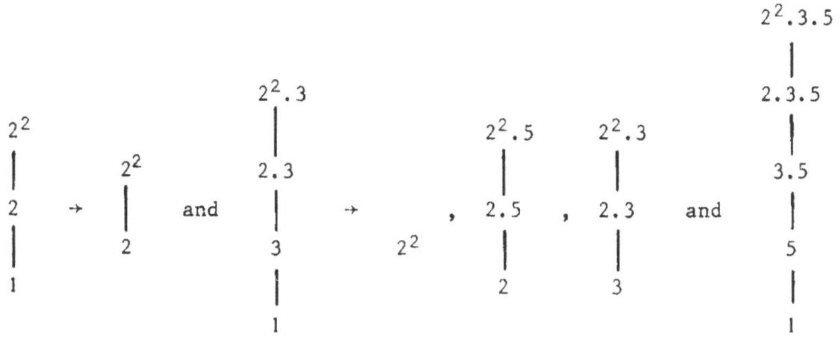

Fig. 24

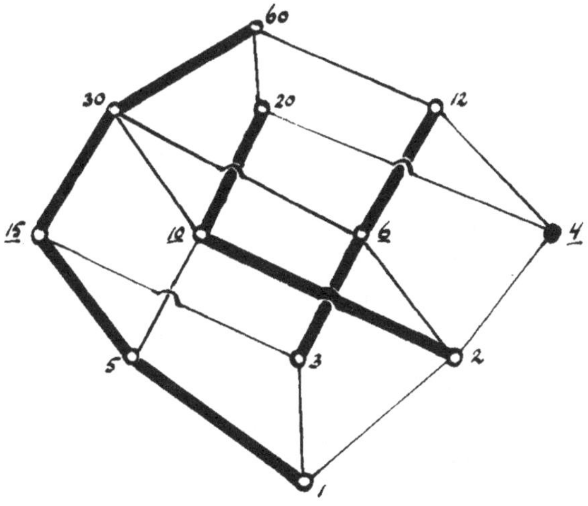

Fig. 25

In figure 25 there are 4 symmetric chains and we find the antichain

$$\{15 = 3.5,\ 10 = 2.5,\ 6 = 2.3,\ 4 = 2.2\}\ .$$

References.

[1] H. Tverberg, On Dilworth's Decomposition Theorem for Partially Ordered Sets, J. Comb. Theory 3 (1967), 305-306.

[2] L. Mirsky, A Dual of Dilworth's Decomposition Theorem, Am. Math. Monthly 78 (1971), 876-877.

VIII. CONVEX SPACES

The discussion of H. Chapter 8 led to only a few remarks concerning the theorems on convex sets. We present these below. First of all it is worth noting that the statement of H. Theorem 8.1.2 for a point P also holds if P is a convex set.

THEOREM 8.1.1. *Let* C *and* D *be disjoint convex sets in* $\mathbb{R}^n$. *Then*

$$\exists_{\underline{u}\neq\underline{0}}\ \exists_{\alpha\in\mathbb{R}}\ [\forall_{\underline{x}\in C}\ [(\underline{u},\underline{x}) \le \alpha] \wedge \forall_{\underline{y}\in D}\ [(\underline{u},\underline{y}) \ge \alpha]] \ .$$

Proof. Consider the set $C - D := \{\underline{c} - \underline{d} \mid \underline{c} \in C,\ \underline{d} \in D\}$. This is clearly a convex set. Since C and D are disjoint, $\underline{0} \notin C - D$. Hence the result follows from the separation theorem (H. Theorem 8.1.2).

H. Theorem 8.1.5 is the only result of H. § 8.1 which is used in the remaining sections of H. Chapter 8. Therefore, a more direct proof of this theorem seems appropriate. First, we prove an auxiliary result. As usual, if C is a closed convex set in $\mathbb{R}^n$ we define $C^* := \{\underline{y} \in \mathbb{R}^n \mid \forall_{\underline{x}\in C}\ [(\underline{x},\underline{y}) \ge 0]\}$.

LEMMA 8.1.1. *If* C *is a closed convex cone and* $\underline{a} \notin C$, *then there exists a* $\underline{y} \in C^*$, *such that* $(\underline{y},\underline{a}) < 0$.

Proof. Let $\underline{b} \in C$ be such that $|\underline{b}-\underline{a}| = \min_{\underline{x}\in C} |\underline{x}-\underline{a}|$ and define $\underline{y} := \underline{b}-\underline{a}$. Then $\underline{y} \neq \underline{0}$. If $\underline{x}$ is an arbitrary point in C then $\lambda\underline{x} + (1-\mu)\underline{b} \in C$ for $\lambda \ge 0$, $\mu \le 1$. Hence

$$|\underline{y}|^2 = |\underline{b}-\underline{a}|^2 \le |\lambda\underline{x} + (1-\mu)\underline{b} - \underline{a}|^2 = |\lambda\underline{x} - \mu\underline{b} + \underline{y}|^2 \ .$$

If we denote the right-hand side by $F(\lambda,\mu)$ we see that $F(0,0) \le F(\lambda,\mu)$ for $\lambda \ge 0$, $\mu \le 1$. A short calculation shows that

$$\left(\frac{\partial F}{\partial\lambda}\right)_{(0,0)} = 2(\underline{y},\underline{x}) \ , \qquad \left(\frac{\partial F}{\partial\mu}\right)_{(0,0)} = -\ 2(\underline{y},\underline{b}) \ .$$

Hence $(\underline{y},\underline{x}) \ge 0$ for all $\underline{x} \in C$, i.e. $\underline{y} \in C^*$. Furthermore, $(\underline{y},\underline{b}) \le 0$ and hence $(\underline{y},\underline{a}) = (\underline{y},\underline{b}) - (\underline{y},\underline{y}) < 0$.

We now prove H. Theorem 8.1.5:

THEOREM 8.1.2. *If* C *is a closed convex cone, then* $(C^*)^* = C$.

Proof. If $\underline{a} \in C$ then $(\underline{a},\underline{y}) \ge 0$ for all $\underline{y} \in C^*$, i.e. $\underline{a} \in (C^*)^*$. If $\underline{a} \notin C$ then by Lemma 8.1.1 there exists a $\underline{y} \in C^*$ with $(\underline{a},\underline{y}) < 0$, i.e. $\underline{a} \notin (C^*)^*$.

We remark that if C is an arbitrary set in $\mathbb{R}^n$ and C^* again denotes the set of points $\underline{y}$ such that $(\underline{x},\underline{y}) \ge 0$ for every $\underline{x} \in C$, then $(C^*)^*$ is the closure of the convex cone cc(C) (cf. § 5.8).

Finally, we make a comment on H § 8.2. The statement that the cone $C := \{\underline{x}A \mid \underline{x} \geq 0\}$ is closed is not so obvious as is suggested on H.p.71. The statement is easily seen to be true if $A: \mathbb{R}^m \to M$ is a 1-1 linear mapping from the vector space $\mathbb{R}^m$ onto the vector space M. Indeed, such a mapping maps closed sets into closed sets and C is the image of the closed set $\{\underline{x} \in \mathbb{R}^m \mid \underline{x} \geq 0\}$.

The general case can be reduced to this special situation by using the cone version of Caratheodory's theorem (Theorem 5.8.1). To see this, let us note that $C = cc(\underline{a}_1,\ldots,\underline{a}_m)$, the convex closed cone generated by the rowvectors $\underline{a}_1,\ldots,\underline{a}_m$ of the matrix A. According to the foregoing, $cc(\underline{a}_1,\ldots,\underline{a}_m)$ is closed if $\underline{a}_1,\ldots,\underline{a}_m$ are linearly independent, for in that case A is a 1-1 linear mapping from $\mathbb{R}^m$ onto $\mathbb{R}^m A$. Now, according to Theorem 5.8.1, in the general case we have

$$cc(\underline{a}_1,\ldots,\underline{a}_m) = \cup(cc(\underline{a}_{i_1},\ldots,\underline{a}_{i_k})) ,$$

where the union is taken over all sequences of indices $i_1,\ldots,i_k$ such that $\underline{a}_{i_1},\ldots,\underline{a}_{i_k}$ are independent. Since there are only finitely many such sequences, we see that $cc(\underline{a}_1,\ldots,\underline{a}_m)$ is closed.

IX. DE BRUIJN SEQUENCES

9.1. *The number of De Bruijn sequences.*

We shall consider a generalization of the cycles studied in H. § 9.1. Instead of symbols 0 and 1 we consider sequences $(a_k)_{k=1}^{\sigma^n}$ of symbols $1,2,\dots,\sigma$ of length σ^n such that (if we consider the symbols in circular order) every possible n-tuple occurs as a subsequence $a_i,a_{i+1},\dots,a_{i+n-1}$. Two such sequences, again called De Bruijn sequences, will be considered distinct if it is not possible to transform one into the other by a circular permutation. We shall count these, using a method different from the method presented in H. § 9.3. The method also applies for $\sigma = 2$ (cf. H. Theorem 9.3.2.Corollary). We consider this method easier to comprehend and it has the additional advantage of yielding an algorithm for constructing such sequences.
The proof depends on a theorem of N.G. de Bruijn and T. van Aardenne-Ehrenfest (cf. [1]) and a theorem of W.T. Tutte (cf. [2]). Once these theorems have been established the only thing that has to be done is the computation of a determinant. We start with some definitions.
An *arborescence* with root P is a directed graph such that for each vertex $Q \neq P$ there is a unique directed path from Q to P (i.e. the graph is a tree). (See figure 26.)

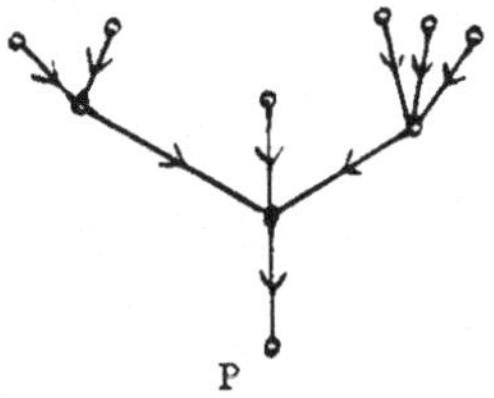

Fig. 26

If G is a directed graph then an arborescence in which all vertices of G occur is called a *spanning* arborescence of G.
Let G be a directed graph on the vertices $P_1,P_2,\dots,P_n$. If, for each vertex P_i, the number of edges starting from P_i is the same as the number of edges pointing towards P_i (we shall denote this number by r_i) then the graph is called *Eulerian*.

THEOREM 9.1.1. *Let* G *be an Eulerian graph on the vertices* $P_1,P_2,\dots,P_n$. *Let* r_i *be the number of edges starting from* P_i. *Let* A_i *be the number of spanning arborescences with root* P_i. *Then the number of distinct directed cycles in* G *using every edge once is*

$$(9.1.1) \qquad A_i \prod_{j=1}^{n} (r_j - 1)!$$

Proof. The proof is nearly the same as for H. Theorem 9.3.1. Consider a spanning arborescence (w.l.o.g. with root P_1). Fix an edge pointing away from P_1 as the first edge in the cycle. Number the other edges pointing away from P_1 arbitrarily. Also number the edges pointing away from P_i $(i = 2,3,\dots,n)$ in an arbitrary way but such that the edge of the spanning arborescence has number r_i. This numbering can be done in $\prod_{j=1}^{n} (r_j - 1)!$ ways. Starting at P_1 and the first edge, we follow a path through G defined as follows. On arrival at P_i leave by the unused edge with lowest number. If, during this process, we arrive at P_i $(i \neq 1)$ then the number of arrivals at P_i exceeds the number of departures by one. Therefore the process terminates in P_1. When this happens, every edge pointing towards P_1 has been used. This means that the points of the spanning arborescence at distance 1 from the root have had all their edges used. By induction it easily follows that all edges of the spanning arborescence and hence all edges of G have been used. Since it is also clear that we can reverse the process to reconstruct a spanning arborescence from a complete cycle, the theorem is proved.

COROLLARY. *In an Eulerian graph* G *on* $P_1, P_2, \dots, P_n$ *the number of spanning arborescences with root* P_i *does not depend on* i.

For a directed graph G on $P_1, P_2, \dots, P_n$ without multiple edges we define the *adjacency matrix* $\mathfrak{A} := [a_{ij}]$ by

$$(9.1.2) \qquad a_{ij} := \begin{cases} 1 \text{ if there is an edge } P_i \to P_j , \\ 0 \text{ otherwise.} \end{cases}$$

Furthermore we associate with G the matrix $\mathfrak{M} := [m_{ij}]$ defined by

$$(9.1.3) \qquad m_{ij} := \begin{cases} \text{the number of edges, not counting loops, pointing away from } P_i \text{ if } j = i , \\ - a_{ij} \text{ if } j \neq i . \end{cases}$$

THEOREM 9.1.2. *Let* G *be a directed graph on* $P_1, P_2, \dots, P_n$ *and let* $\mathfrak{M}$ *be defined by* (9.1.3). *Then the number* A_i *of spanning arborescences of* G *with root* P_i *is the minor of* m_{ii} *in* $\mathfrak{M}$.

Proof. The minor of m_{ii} is a linear function of the rows of $\mathfrak{M}$. If we consider an edge $P_i \to P_j$, then the number A_i is equal to the sum of the number of spanning arborescences with, respectively without, edge $P_i \to P_j$. This means A_i is also a linear function of the rows of $\mathfrak{M}$. Hence it is sufficient to consider a graph G for which the matrix $\mathfrak{M}$ has one row of zeros (say the first row, $i = 1$) and a 1 in all the other diagonal positions $(i > 1)$. If the graph G has a cycle, say $P_2 \to P_3$, $P_3 \to P_4, \dots, P_k \to P_2$, then the sum of rows 2 to k is $\underline{0}$ and hence the minor of m_{11} is 0. Since in this case G has $n - 1$ edges and a cycle it contains no spanning arborescence. It remains to show that if G is a (spanning) arborescence then the minor of m_{11} is 1.

This is immediately seen by induction on n.
We now are in a position to count the number of De Bruijn sequences of length σ^n with symbols $1,2,\ldots,\sigma$. In the same way as in H. Chapter 9 we associate with this problem a graph G on the "points" $(a_1,a_2,\ldots,a_{n-1})$, i.e. all $(n-1)$-tuples from the symbols $1,2,\ldots,\sigma$. There is an edge $(a_1,a_2,\ldots,a_{n-1}) \to (a_2,a_3,\ldots,a_{n-1},x)$ for $x = 1,2,\ldots,\sigma$. Clearly G is an Eulerian graph with $r_i = \sigma$ for all vertices P_i. Here we number the vertices $P_1,P_2,\ldots,P_{\sigma^{n-1}}$ according to the lexicographic ordering of the $(n-1)$-tuples. The number of distinct directed cycles in G using every edge once is equal to the number of distinct De Bruijn sequences of length σ^{n-1}. Now consider the matrix $\mathfrak{M}(n,\sigma) := [c_{ij}]$ of order σ^{n-1} defined as follows:

$$c_{ij} := \begin{cases} 1 \text{ if } \sigma(i-1) (\mathrm{mod}\ \sigma^{n-1}) < j \le \sigma i\ (\mathrm{mod}\ \sigma^{n-1}), & 1 \le i \le \sigma^{n-1}, \\ 0 \text{ otherwise.} \end{cases}$$

Then the matrix $\mathfrak{M}$ corresponding to G, as defined in (9.1.3) is $\sigma I - \mathfrak{M}(n,\sigma)$.

For the proof of the main theorem we need the characteristic polynomial of $\mathfrak{M}(n,\sigma)$, i.e. $\det(\lambda I - \mathfrak{M}(n,\sigma))$. To calculate this determinant we perform the obvious row and column operations: The block consisting of the first σ^{n-2} rows is subtracted from subsequent blocks of σ^{n-2} rows with as result that all rows with number $> \sigma^{n-2}$ now have one term λ (on the diagonal) and one term $-\lambda$. We eliminate the $-\lambda$'s by column operations. The result is

$$(9.1.4) \qquad \det(\lambda I - \mathfrak{M}(n,\sigma)) = \lambda^{(\sigma-1)\sigma^{n-2}} \det(\lambda I - \mathfrak{M}(n-1,\sigma)) .$$

From (9.1.4) we find

$$(9.1.5) \qquad \det(\lambda I - \mathfrak{M}(n,\sigma)) = \lambda^{\sigma^{n-1}-1}(\lambda - \sigma)$$

since $\mathfrak{M}(2,\sigma) = J$.

We now come to the proof of the main theorem.

THEOREM 9.1.3. *The number of distinct De Bruijn sequences of length* σ^n *on the symbols* $1,2,\ldots,\sigma$ *is* $\sigma^{-n}(\sigma!)^{\sigma^{n-1}}$.

Proof. By the preceding introduction and Theorems 9.1.1 and 9.1.2 we find that the required number is

$$(9.1.6) \qquad A_i((\sigma-1)!)^{\sigma^{n-1}}$$

where A_i is the minor of the i-th diagonal element of $\sigma I - \mathfrak{M}(n,\sigma)$. This minor does not depend on i $(i = 1,2,\ldots,\sigma^{n-1})$. For any matrix B the sum of the minors of the diagonal elements is equal to the coefficient of λ in the polynomial $\det(\lambda I + B)$. From (9.1.5) we therefore have

$$\sigma^{n-1} A_i = \text{coefficient of } \lambda \text{ in } (\lambda+\sigma)^{\sigma^{n-1}-1}\lambda ,$$

i.e.

$$A_i = \sigma^{\sigma^{n-1}-n} . \tag{9.1.7}$$

From (9.1.6) and (9.1.7) the theorem follows. (For generalizations see [3], [4].)

The proof using Theorem 9.1.1 yields an easy algorithm for constructing De Bruijn sequences. Consider once again the graph G described above. The edges $(a_1,a_2,\dots,a_{n-1}) \to (a_2,\dots,a_{n-1},1)$ for $(a_1,a_2,\dots,a_{n-1}) \neq (1,1,\dots,1)$ clearly form a spanning arborescence with root $(1,1,\dots,1)$. Let the loop on this point be the distinguished edge starting the complete circuit in G. As in the proof of Theorem 9.1.1 we number the edges leaving $(a_1,a_2,\dots,a_{n-1})$ as follows: $(a_1,a_2,\dots,a_{n-1}) \to (a_2,a_3,\dots,a_{n-1},x)$ has number $\sigma - x + 1$. The circuit described in the proof yields a De Bruijn sequence which can also be described by the following algorithm.

ALGORITHM. The sequence $a_1,a_2,\dots,a_{\sigma^{n-1}}$ is defined by $a_1 := a_2 := \dots := a_n := 1$, and for $k > n$ a_k is defined to be the largest integer in $\{1,2,\dots,\sigma\}$ such that the sequence $(a_{k-n+1},\dots,a_{k-1},a_k)$ has not occurred in $(a_1,\dots,a_{k-1})$ as a (consecutive) subsequence. An example for $\sigma = n = 3$ is

1 1 1 3 3 3 2 3 3 1 3 2 2 3 2 1 3 1 2 3 1 1 2 2 2 1 2 .

Note that if we make the cyclic shift such that the sequence starts at a_{n+1} and ends with $1,1,\dots,1$, then the resulting De Bruijn sequence is the last sequence in the lexicographic ordering of all De Bruijn sequences of this length. For $\sigma = 2$ the resulting sequence is known as the *Ford sequence* (cf. [5], [6]).

The proof of theorem 9.1.1 shows that if we use the same spanning arborescence as above, but no longer insist on the numbering we defined for the algorithm, then this method leads to $((\sigma-1)!)^{\sigma^{n-1}}$ distinct De Bruijn sequences.

Recently a different algorithm for constructing De Bruijn sequences was described by E. Roth (cf. [7]). We shall describe it graphically in accordance with the point of view in this section. Once again consider the graph G we used above. Let $a_1,a_2,\dots,a_n$ be a sequence of integers chosen from $\{1,2,\dots,\sigma\}$ which is periodic with a period d where $d \mid n$ and $d = n$ if the sequence is not periodic. Then $(a_1,a_2,\dots,a_{n-1}) \to$ $\to (a_2,\dots,a_n) \to (a_3,\dots,a_n,a_1) \to \dots \to (a_d,a_{d+1},\dots,a_{d-2}) \to (a_1,a_2,\dots,a_{n-1})$ is a circuit in G. The graph G is the union of (edge) disjoint circuits of this type. By (H.2.1.18) and (H.2.1.19) the number of circuits is $\sum_{d|n} \frac{1}{d} \sum_{k|d} \mu(d/k)\sigma^k$. Of course two such circuits can have vertices of G in common. In figure 27 we give an example for

$\sigma = 2$, $n = 4$.

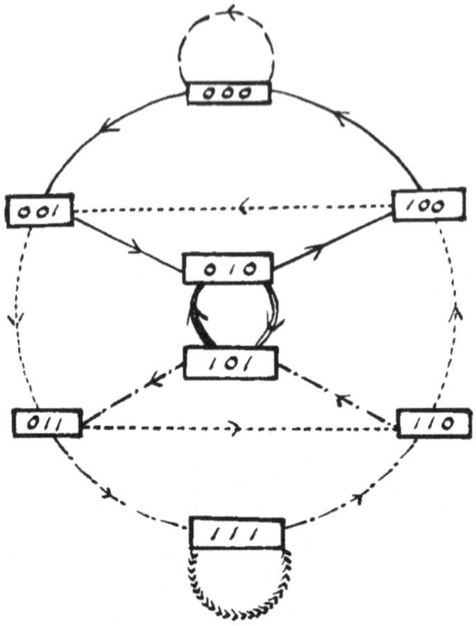

Fig. 27

The n-tuples (0,0,0,1), (1,0,0,1), (0,1,1,1) correspond to circuits with 4 edges, the 4-tuple (0,1,0,1) to a circuit with 2 edges, and (0,0,0,0) and (1,1,1,1) correspond to the 2 loops.

The algorithm works as follows. Take any of the circuits to start with. Take a point on this circuit which is also on another circuit and combine the two to make a new circuit. On this new circuit take a point which is on a circuit which has not been used and insert this circuit, etc. An example for figure 27 is:

(a) $(0,0,0) \to (0,0,0)$,

(b) $(0,0,0) \to (0,0,0) \to (0,0,1) \to (0,1,0) \to (1,0,0) \to (0,0,0)$,

(c) $(0,0,0) \to (0,0,0) \to (0,0,1) \to (0,1,1) \to (1,1,0) \to (1,0,0) \to (0,0,1) \to (0,1,0) \to (1,0,0) \to (0,0,0)$,

etc.

After 6 steps the De Bruijn cycle is complete. Note that after each step of the algorithm we have a sequence with the property that if an $(n-1)$-tuple $(a_1, a_2, \ldots, a_{n-1})$ is a subsequence then all cyclic shifts of this $(n-1)$-tuple are also subsequences. This algorithm also has the property that a De Bruijn sequence of length $(\sigma-1)^n$ on the symbols $1, 2, \ldots, \sigma-1$ can be used as a starting point in the construction of a De Bruijn sequence of length σ^n on the symbols $1, 2, \ldots, \sigma$. It is not difficult to construct examples of De Bruijn sequences which cannot be produced by Roth's algorithm. An example can be found from figure 27: 0000111100101101. If we try to reverse the algorithm, only the loops at 0 0 0 and 1 1 1 can be removed. Even for $n = 2$ there is already a counterexample when $\sigma = 4$. In the corresponding graph G the circuit C: $1 \to 2 \to 3 \to 1 \to 4 \to 2 \to 1 \to 3 \to 2 \to 4$ contains all edges except the pair $3 \to 4$, $4 \to 3$ and the four loops on 1, 2, 3, 4. Hence if we start Roth's algorithm with C it yields a De Bruijn sequence but it is not possible to reduce C. These two counterexamples of length 16 are the shortest ones possible. Concerning this algorithm we state the following interesting open problem:

PROBLEM. How many distinct De Bruijn sequences of length σ^n on $1,2,\ldots,\sigma$ can be obtained by Roth's algorithm?

By hand calculation we found that for the case $\sigma = 4$, $n = 2$, mentioned above, the answer is 13/16 of the total number of De Bruijn sequences.

For the sake of completeness we mention that it was shown by A. Lempel [8] that for $k \le \sigma^n$ there exists a circular sequence $a_1,a_2,\ldots,a_k$ such that all subsequences $a_i,a_{i+1},\ldots,a_{i+n-1}$ are different. In terms of the graph G which we have used this means that there are circuits in G of length k for all $k \le \sigma^n$.

9.2. *Shift register sequences.*

A shift register is a sequence of memory elements represented by the sequences $x_1,x_2,\ldots,x_n$ in figure 28. The contents of these elements are 0's and 1's (in the

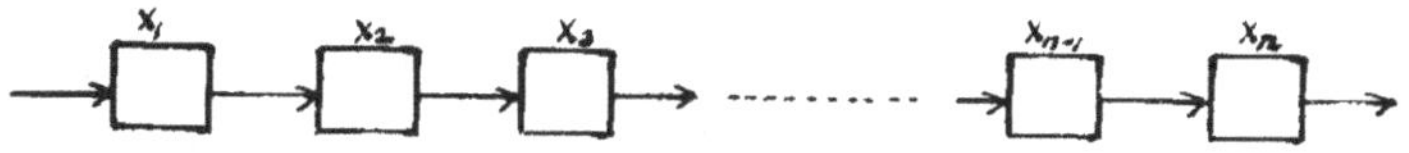

Fig. 28

binary case to which we restrict our attention). At periodic intervals determined by a master clock the contents of x_i are transferred to x_{i+1} $(i = 1,2,\ldots,n-1)$ and the contents of x_n leave the register. There are several possible mechanisms governing what comes into x_1 at the periodic intervals. One example is the *feadback shift register* (FSR) of figure 29:

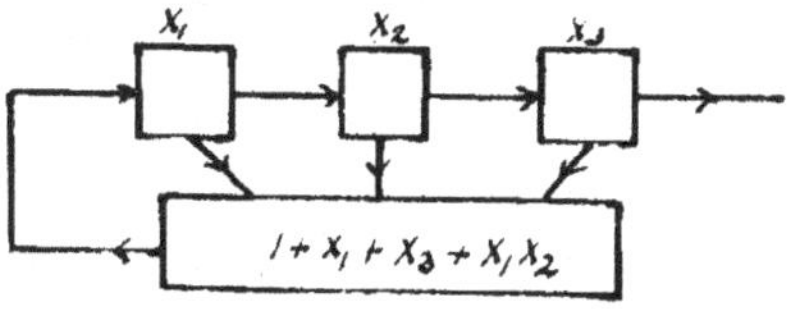

Fig. 29

In this example at each shift the value of $1 + x_1 + x_3 + x_1x_2$ (arithmetic in GF(2)) is calculated and transferred into x_1. The sequence of 0's and 1's leaving the register is obviously periodic. In the case of figure 29 the period is 8. If we start with $x_1 = x_2 = x_3 = 0$ then the sequence is $\rightarrow, 1, 1, 1, 0, 1, 0, 0, 0 \rightarrow$.
In general for n memory elements the period is $\leq 2^n$. If the period is maximal, i.e. 2^n, then a period (interpreted as a circular sequence!) is a De Bruijn sequence. The example of figure 29 shows a De Bruijn sequence of length 8 generated by an FSR. Such sequences are sometimes called *maximal length feadback shift register sequences*.

We give another example of a sequence generated by a shift register. Consider figure 30.

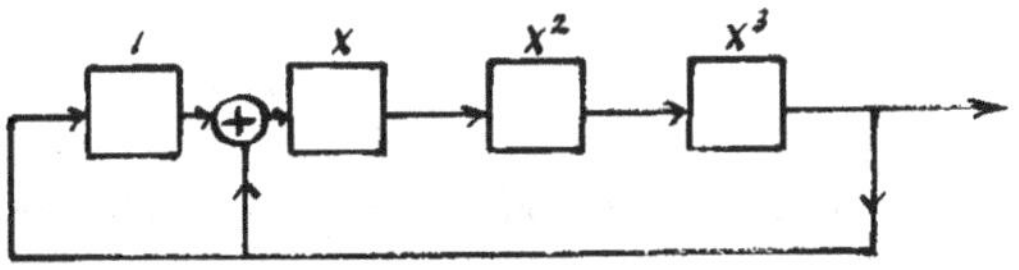

Fig. 30

The contents a_0, a_1, a_2, a_3 of the four elements of the shift register are considered as coefficients of a polynomial $a_0 + a_1x + a_2x^2 + a_3x^3$. When a shift occurs the coefficient of x^3 is transferred into the first square and simultaneously with the shift described above it is added (mod 2) to the previous contents of the first element before this enters the second element. In formula we have the shift from $c_1(x) :=$ $:= a_0 + a_1x + a_2x^2 + a_3x^3$ to $c_2(x) := a_3 + (a_0 + a_3)x + a_1x^2 + a_2x^3$. Here $c_2(x) =$ $= xc_1(x) \bmod(x^4 + x + 1)$. There are two possibilities. If initially the register contains only 0's, this will remain the case. Otherwise the register will subsequently contain the other 15 possible states. The sequence leaving the register is periodic with period 15. In this case a period is a De Bruijn sequence in which one zero has been omitted from the occurring 4-tuple: 0,0,0,0. We call this a *shortened* De Bruijn sequence. If the register originally contains 1,0,0,0 (i.e. the polynomial 1) then the following states of the register are $x, x^2, \ldots, x^{14} \bmod(x^4 + x + 1)$. This process can also be described as follows. Let x be a primitive element of the field $GF(2^4)$ with minimal polynomial $x^4 + x + 1$ (a primitive factor of $x^{15} - 1$). Then the non-zero elements of $GF(2^4)$ form the cyclic group generated by x. Each power of x can be expressed as a polynomial of degree ≤ 3 in x. The successive states of the shift register are these polynomials.

Let us consider the general case. Consider a primitive polynomial h(x) of degree k over GF(2), i.e. a polynomial such that $n = 2^k - 1$ is the smallest positve integer such that $x^n - 1 \equiv 0 \pmod{h(x)}$. A shift register with k elements behaves as in figure 30, now with the shift $c(x) \mapsto xc(x) \pmod{h(x)}$. We start with $1,0,0,\ldots,0$. The elements leaving the shift register are the subsequent coefficients of x^{k-1} in x^i (mod h(x)), $i = 0,1,2,\ldots$. If i and j are such that the coefficients of x^{k-1} in

$x^i, x^{i+1}, \ldots, x^{i+k-1}$ (mod h(x)), respectively $x^j, x^{j+1}, \ldots, x^{j+k-1}$ (mod h(x)) are the same k-tuples, then the polynomials $x^j - x^i$, $x^{j+1} - x^{i+1}$, ..., $x^{j+k-1} - x^{i+k-1}$ (mod h(x)) are linearly dependent over GF(2). Therefore there are coefficients $c_0, c_1, \ldots, c_{k-1}$ such that

$$\sum_{\nu=0}^{k-1} c_\nu (x^{j+\nu} - x^{i+\nu}) \equiv 0 \pmod{h(x)} ,$$

i.e.

$$(x^j - x^i) \sum_{\nu=0}^{k-1} c_\nu x^\nu \equiv 0 \pmod{h(x)}$$

and since h(x) is a primitive polynomial this implies $j \equiv i \pmod{2^k - 1}$. This proves that the sequence leaving the shift register is periodic with period $2^k - 1$. Now, somewhere in this sequence, $k - 1$ consecutive 0's occur. If we insert one more 0 here, the resulting sequence is a De Bruijn sequence.

Since we start with 1,0,0,...,0 in the register, we see that the elements leaving the register are the coefficients of $x^{n-1}, x^{n-2}, \ldots, 1$ which result when x^n is divided by h(x). Since we have a period when the register is back to 1,0,0,...,0 after n steps we see that the sequence leaving the register, when interpreted as a polynomial, is $(x^n - 1)/h(x)$.

We now consider the ring R of polynomials GF(2)[x] mod $x^n - 1$ as an n-dimensional vector space over GF(2). If we define $g(x) := (x^n - 1)/h(x)$ then the multiples of g(x) form a subspace C of dimension k in R. If $a_0 + a_1 x + \ldots + a_{n-1} x^{n-1}$ is in C then $x(a_0 + a_1 x + \ldots + a_{n-1} x^{n-1}) \bmod x^n - 1$, i.e. $a_{n-1} + a_0 x + a_1 x^2 + \ldots + a_{n-2} x^{n-1}$ is also in C. Such a subspace is called a *cyclic code*. Since C contains 0 and the $n = 2^k - 1$ cyclic shifts of g(x) and C has dimension k, we see that these are the only elements of C. We formulate this as follows:

THEOREM 9.2.1. *If* h(x) *is a primitive polynomial of degree* k *over* GF(2), $n := 2^k - 1$, *and* $g(x) := (x^n - 1)/h(x)$ *then the cyclic shifts* $g(x), xg(x), \ldots, x^{n-1}g(x) \pmod{x^n - 1}$ *and* 0 *form an additive group* C.

We remark that if we compare any two different polynomials in C, then the number of places where the coefficients differ is equal to the number of nonzero coefficients of g(x), i.e. it is constant. Such a code C is called an *equidistant code*. (In coding theory the (Hamming-) distance of two vectors in n-space is the number of coordinate places where they differ.)

9.3. *Randomness properties of De Bruijn sequences.*

Let $a_1, a_2, \ldots, a_N$ be a sequence of "random numbers" where $a_i \in \{1,2,\ldots,\sigma\}$, $(i = 1,2,\ldots,N)$. In such a sequence a subsequence $a_i, a_{i+1}, \ldots, a_{i+k-1}$ such that $a_i = a_{i+1} = \ldots = a_{i+k-1}$ and $a_{i-1} \neq a_i$, $a_{i+k} \neq a_{i+k-1}$ is called a *run* of length k. The probability that a_i is a run of length 1 is clearly $(\frac{\sigma-1}{\sigma})^2$. Hence if N is large we expect the sequence to contain approximately $(\frac{\sigma-1}{\sigma})^2 N$ runs of length 1. In the same way we see that if k is small compared to N we may expect approximately $(\sigma-1)^2 \sigma^{-k-1} N$ runs of length k. We also observe that we expect each of the σ elements to occur approximately $\sigma^{-1} N$ times in the sequence.

We now show that every De Bruijn sequence has these same properties. If $a_1, a_2, \ldots, a_N$, where $N = \sigma^n$, is a De Bruijn sequence then obviously every element occurs σ^{n-1} times in the sequence. By definition there are σ runs of length n in the sequence. If $a_i = a_{i+1} = \ldots = a_{i+n-1} = a$ is the run of n a's and $a_{i+n} = b$ is the next element, then the definition of a De Bruijn sequence implies that a run of $n-1$ a's (if any) is followed by an element c with $c \neq a$, $c \neq b$. Hence there are $\sigma(\sigma-2)$ runs of length $n-1$. Now let $k < n-1$. Every n-tuple $c_1, c_2, \ldots, c_n$ with $c_1 \neq a$, $c_2 = c_3 = \ldots = c_{k+1} = a$, $c_{k+2} \neq a$, occurs once in the De Bruijn sequence. It follows that there are $(\sigma-1)^2 \sigma^{n-k-1} = (\sigma-1)^2 \sigma^{-k-1} N$ runs of length k. Therefore both for runs of length $k < n-1$ and according to occurrence of the different symbols a De Bruijn sequence behaves as a random sequence.

We now define a special type of sequence with randomness properties called a *pseudo-noise* sequence (PN-sequence).

DEFINITION. *Let* $a_1, a_2, \ldots, a_N$ *be a circular sequence (i.e.* $a_{i+N} := a_i$ *for all* i*) of* +1*'s and* -1*'s. If*

(i) $\left|\sum_{i=1}^{N} a_i\right| \leq 1$,

(ii) *the total number of runs is* 2^m *and for* $k = 1,2,\ldots,m-1$ *there are* 2^{m-k-1} *runs of* k +1*'s and* 2^{m-k-1} *runs of* k -1*'s and there is one run of length* m,

(iii) $\sum_{i=1}^{N} a_i a_{i+j} = \begin{cases} N \text{ if } j = 0 , \\ \alpha \text{ if } j \neq 0 , \end{cases}$

then the sequence is called a pseudo-noise sequence.

We have shown above that if we take $\sigma = 2$, $N = 2^n - 1$ then a shortened De Bruijn sequence of length N (where we now use +1 and -1 for the two symbols) has the properties (i) and (ii) of a PN-sequence. We shall now prove that a shift register of the type of figure 30 using a primitive polynomial h(x) of degree k $(N = 2^k - 1)$ produces a shortened De Bruijn sequence which is a PN-sequence. Here we map 0 and 1 into +1 and -1. To show this we apply Theorem 9.2.1. This theorem states that the sequences $a_1, a_2, \ldots, a_N$ and $a_{1+j}, a_{2+j}, \ldots, a_{N+j}$ $(j \neq 0)$ both have $2^{k-1} - 1$ elements +1 and 2^{k-1} elements -1 and that the sequences coincide in $2^{k-1} - 1$ places. It follows that $\sum_{i=1}^{N} a_i a_{i+j} = -1$, i.e. (iii) is satisfied with $\alpha = -1$.

We make one final observation in connection with Chapter 14. Let $a_1, a_2, \ldots, a_N$ $(N = 2^k - 1)$ be the shortened De Bruijn sequence of +1's and -1's discussed above and define the matrix H as follows:

$$(9.3.1) \qquad H := \begin{pmatrix} 1 & 1 & 1 & \cdots & \cdots & 1 \\ 1 & a_1 & a_2 & \cdots & \cdots & a_N \\ 1 & a_N & a_1 & a_2 & \cdots & a_{N-1} \\ \vdots & \cdots & \cdots & \cdots & \cdots & \vdots \\ 1 & a_3 & a_4 & \cdots & a_1 & a_2 \\ 1 & a_2 & a_3 & \cdots & a_N & a_1 \end{pmatrix}.$$

Then $HH^T = (N+1)I$, i.e. H is a Hadamard matrix which is a circulant bordered by a row and column of 1's.

The reader who wishes to know more about the subject discussed in sections 9.2 and 9.3 is referred to S.W. Golomb's book [9].

References:

[1] N.G. de Bruijn and T. van Aardenne-Ehrenfest, Circuits and Trees in Oriented Linear Graphs, Simon Stevin 28 (1951), 203-217.

[2] W.T. Tutte, The Dissection of Equilateral Triangles into Equilateral Triangles, Proc. Cambr. Phil. Soc. 44 (1948), 463-482.

[3] R. Dawson and I.J. Good, Exact Markov Probabilities from Oriented Linear Graphs, Ann. Math. Stat. 28 (1957), 946-956.

[4] D.E. Knuth, Oriented Subtrees of an Arc Digraph, J. Comb. Yheory 3 (1967), 309-314.

[5] L.R. Ford, Jr., A Cyclic Arrangement of M-tuples, Report P-1071, Rand Corporation, Santa Monica, Cal. (1957).

[6] H. Fredricksen, The Lexicographically Least De Bruijn Cycle, J. Comb. Theory 9 (1970), 1-5.

[7] E. Roth, Permutations Arranged around a Circle, Amer. Math. Monthly (1971), 990-992.

[8] A. Lempel, m-ary Closed Sequences, J. Comb. Theory 10 (1971), 253-258.

[9] S.W. Golomb, Shift Register Sequences, Holden-Day, Inc., San Francisco (1967).

X. BLOCK DESIGNS

10.1. *Block designs.*

In the following we shall use the notation

(10.1.1) $B(n,k,\lambda) := (k-\lambda)I + \lambda J$ (of order n) .

A block design on v objects with b blocks of k distinct objects, each object occurring in r different blocks and each pair of distinct objects ocurring in λ blocks, will be denoted by $BD(v,k\,;b,r,\lambda)$. Here the first two parameters are connected with the rows of the $b \times v$ incidence matrix A of the design and the other three concern properties of the columns of A. (We prefer representing blocks as rows.) If the design is symmetric, i.e. $b = v$, then A satisfies

(10.1.2) $AA^T = A^TA = B(v,k,\lambda)$

and

(10.1.3) $AJ = JA = kJ$.

In H. Theorem 10.2.3 it is shown that if A is nonsingular of order v and A satisfies one of the two equations of (10.1.2) and one of the two of (10.1.3) then A satisfies all four of the equations. Furthermore, H. Theorem 16.4.2 states that if A is a $v \times v$ matrix of integers which satisfies (10.1.2) and if $k(k-1) = \lambda(v-1)$ then A or $-A$ is the incidence matrix of a block design. We now consider some other results on matrix equations of this same type which have many combinatorial applications.
The first generalization is due to W.G. Bridges and H.J. Ryser (cf. [1]).

THEOREM 10.1.1. *Let* X *and* Y *be nonnegative integral matrices of order* $n > 1$ *such that*

(10.1.4) $XY = B(n,k,\lambda)$.

Let $k \neq \lambda$ *and let the integers* k *and* λ *be relatively prime. Then there exist positive integers* r *and* s *with* $rs = k + (n-1)\lambda$ *and*

(10.1.5) $XJ = JX = rJ$

(10.1.6) $YJ = JY = sJ$,

(10.1.7) $XY = YX$.

Proof. We use the abbreviation B for $B(n,k,\lambda)$ and we define $\Delta := k + (n-1)\lambda$. The matrix B is nonsingular with

(10.1.8) $B^{-1} = (k-\lambda)^{-1}\, I - (\Delta(k-\lambda))^{-1}\, \lambda J$.

Now (10.1.4) implies

(10.1.9) $X(YB^{-1}) = (YB^{-1})X = I$.

In the second of these equations we substitute (10.1.8). This yields

(10.1.10) $$YX = (k-\lambda)I + \lambda\Delta^{-1}YJX .$$

If we denote by f_i the sum of the i-th row of Y and by e_j the sum of the j-th column of X then

(10.1.11) $$YJX = [f_i e_j] .$$

Now $(k,\lambda) = 1$ implies $(\Delta,\lambda) = 1$ and since the elements of YX are integers we see that $\Delta \mid f_i e_j$ $(i,j = 1,2,\ldots,n)$. In the trivial case $\lambda = 0$ this is also true. Since X and Y are nonsingular we have $f_i > 0$, $e_j > 0$ $(i,j = 1,2,\ldots,n)$ and therefore $f_i e_j \geq \Delta$. By (10.1.4) and (10.1.10) we have

$$nk = tr(XY) = tr(YX) = n(k-\lambda) + \lambda\Delta^{-1}\sum_{i=1}^{n} e_i f_i .$$

i.e.

(10.1.12) $$\sum_{i=1}^{n} e_i f_i = n\Delta .$$

But then $e_i f_i = \Delta$ $(i = 1,2,\ldots,n)$. Therefore $e_i \leq e_j$ $(i,j = 1,2,\ldots,n)$, i.e. all e_i are equal, say r, and hence all f_j are equal, say s, and $rs = \Delta$. For (10.1.10) we thus have

$$YX = (k-\lambda)I + \lambda J = B(n,k,\lambda) .$$

If we now interchange X and Y the proof is complete.

Remark. It is not difficult to give examples which show that the restriction $(k,\lambda) = 1$ cannot be dropped (cf. [1]). Although there is a great analogy with the two theorems mentioned earlier, we cannot derive them from Theorem 10.1.1. For interesting applications to combinatorial designs we refer to the original paper.

A second generalization of the equations (10.1.1), (10.1.4), also due to H.J. Ryser (cf. [2]), replaces I by a diagonal matrix and the offdiagonal elements λ in position (i,j) by $\sqrt{\lambda_i}\sqrt{\lambda_j}$ where λ_i $(i = 1,2,\ldots,n)$ are nonnegative numbers. Then (10.1.1) is a special case of the theorem.

THEOREM 10.1.2. *Let* $X := [x_{ij}]$ *and* $Y := [y_{ij}]$ *be real matrices of order* n *that satisfy the matrix equation*

(10.1.13) $$XY = D + [\sqrt{\lambda_i}\sqrt{\lambda_j}] ,$$

where D *denotes the diagonal matrix* $\mathrm{diag}[r_1 - \lambda_1,\ldots,r_n - \lambda_n]$, *and the numbers* $r_i - \lambda_i$ *and* λ_i *are positive and nonnegative, respectively. Then*

(10.1.14) $$YD^{-1}X = I + w[y_i x_j] ,$$

where the numbers w, y_i *and* x_j *are defined by*

(10.1.15) $$w := 1 + \sum_{i=1}^{n} \lambda_i (r_i - \lambda_i)^{-1} ,$$

(10.1.16) $$wy_i := \sum_{j=1}^{n} \sqrt{\lambda_j} (r_j - \lambda_j)^{-1} y_{ij} ,$$

(10.1.17) $$wx_j := \sum_{i=1}^{n} \sqrt{\lambda_i} (r_i - \lambda_i)^{-1} x_{ij} .$$

For the proof we refer to [2]. The idea is similar to the proof of Theorem 10.1.1. Since XY is nonsingular one can multiply (10.1.13) on the left by X^{-1} and on the right by $D^{-1}X$. It then remains to show that the second term which results on the right-hand side is equal to $w[y_i x_j]$. We shall apply this theorem to block designs in the next paragraph.

For a somewhat different generalization we consider once again the graph associated with the De Bruijn sequences of length σ^3 on the symbols $1,2,\ldots,\sigma$. This graph has the property that for every pair of points P_i, P_j there is exactly one directed path of length 2 from P_i to P_j. If A is the adjacency matrix of this graph then this means that $A^2 = J$. It is natural to raise the question whether other such graphs exist. The following more general question was considered by H.J. Ryser: Let A be a (0,1)-matrix of order n for which

(10.1.18) $$A^2 = D + \lambda J$$

where D is a diagonal matrix. What can be said about A? To illustrate his methods we treat the equation $A^2 = J$ and then state Ryser's result without proof (cf. [3]). Let A be a (0,1)-matrix with $A^2 = J$. Then we find

(10.1.19) $$A^3 = AJ = JA .$$

Define $e := \sum_{i=1}^{n} \sum_{j=1}^{n} a_{ij}$. Then from (10.1.19) it follows that

$$eJ = JAJ = nJA ,$$

i.e.

$$JA = AJ = \frac{e}{n} J .$$

Therefore

$$nJ = J^2 = JA^2 = (JA)A = \frac{e}{n} JA = \frac{e^2}{n^2} J ,$$

i.e. $e^2 = n^3$ and therefore n is a square, say $n = c^2$, and

(10.1.20) $\quad AJ = cJ$.

The same result follows easily from a consideration of the eigenvalues of A^2. From $A^2 = J$ and $n = c^2$ it follows that $tr(A) = c$. The question of classifying all solutions of the equation $A^2 = J$ is open. Even for small values of c, e.g. $c = 3$, there are many inequivalent solutions. For $c = 3$ (i.e. $n = 9$) there are 6 nonisomorphic graphs with incidence matrix A for which $A^2 = J$ (cf. [4]). We now state without proof Ryser's generalization.

THEOREM 10.1.3. *Let* A *be a* (0,1) *matrix of order* $n > 1$ *which satisfies the equation*

$$A^2 = D + \lambda J$$

where D *is a diagonal matrix and* λ *is a positive integer. Then*

$$AJ = JA = cJ$$

$$A^2 = dI + \lambda J \; ,$$

with

$$-\lambda < d = c^2 - \lambda n \le c - \lambda \; ,$$

with the following exceptions:

$$\begin{bmatrix} 0 & 1 & 1 & 1 & 1 \\ 1 & 1 & 1 & 0 & 0 \\ 1 & 0 & 0 & 1 & 1 \\ 1 & 1 & 1 & 0 & 0 \\ 1 & 0 & 0 & 1 & 1 \end{bmatrix}, \quad \begin{bmatrix} 1 & 1 & \cdots & 1 \\ 1 & & & \\ \vdots & & O & \\ 1 & & & \end{bmatrix}, \quad \begin{bmatrix} 0 & 1 & \cdots & 1 \\ 1 & & & \\ \vdots & & Q & \\ 1 & & & \end{bmatrix},$$

Q *a symmetric permutation matrix of order* $n-1$.

We remark that the matrix $\mathfrak{M}(3,\sigma)$ defined in section 9.1 is a solution of the equation $A^2 = J$ for the order $n = \sigma^2$ (then $AJ = JA = \sigma J$).
We remark that Theorem 10.1.3 provides a proof of the so-called "*Friendship Theorem*". This theorem states that, in a party of n persons, if every pair of persons has exactly one common friend, then there is someone in the party who is everyone else's friend. The problem is equivalent to finding a graph G on n points $P_1, P_2, \ldots, P_n$ (without loops or multiple edges) such that for each pair P_i, P_j $(i \ne j)$ there is exactly one k such that (P_i, P_k) and (P_k, P_j) are edges of G. If A is the adjacency matrix of G then we must have $A^2 = D + J$ where D is a diagonal matrix. Since A is symmetric we find from Theorem 10.1.3 (excluding the exceptions) that $A^2 = (c-1)I + J$, $AJ = cJ$ and therefore A is the incidence matrix of a $BD(n,c\,;n,c,1)$ and $n = c^2 - c + 1$. Since G has no loops we must have $tr(A) = 0$. If $c - 1$ is not a square then $tr(A) = c$. If $c - 1 = \alpha^2$ then A has eigenvalues $\pm\alpha$ and one eigenvalue $\alpha^2 + 1$. Then $tr(A) = 0$ implies $\alpha = 1$, i.e. $n = 3$. In this case G is a triangle. The only remaining possibility is

the third exception which corresponds to a graph G consisting of a number of triangles which have a common vertex (corresponding to the "mutual friend").

A generalization of the concept of Hadamard matrix as treated in H. Ch.14 is given in the following definition.

DEFINITION. *A C-matrix of order* v *is a square matrix* C *with diagonal elements* 0 *and all other elements* +1 *or* -1, *which satisfies*

(10.1.21) $$CC^T = (v-1)I .$$

If we multiply some of the rows or columns of C by -1 we obtain another C-matrix. In this way we can transform C into a matrix of the form

$$\begin{bmatrix} 0 & \underline{j}^T \\ \underline{j} & S \end{bmatrix}$$

where $\underline{j}^T := (1,1,\ldots,1)$. Here S is a square matrix of order $v-1$ satisfying

(10.1.22) $$S\underline{j} = \underline{0} , \quad S^TS = SS^T = (v-1)I - J .$$

We shall call such matrices S-matrices. Note that if C is a skew C-matrix then $H := I + C$ is a Hadamard matrix. This idea is used in Paley's construction of Hadamard matrices (cf. H.p.209).
Generalizing these concepts we consider square matrices A of order v with diagonal elements 0 and all other elements +1 or -1, satisfying the matrix equation

(10.1.23) $$AA^T = B(v,v-1,\lambda) ,$$

We shall prove a theorem concerning such matrices due to P. Delsarte, J.-M. Goethals and J.J. Seidel (cf. [5]). Note that if D is a diagonal matrix with diagonal elements +1 and -1, then AD also satisfies (10.1.23). We shall call A_1 and A_2 equivalent if $A_1D = A_2$ where D is a ± 1-diagonal matrix.

THEOREM 10.1.4. *If* A *satisfies* (10.1.23) *and* $\lambda \neq 0$ *then there is a matrix* B *equivalent to* A *satisfying*

(10.1.24) $$BB^T = B^TB = B(v,v-1,\lambda) ,$$

(10.1.25) $$BJ = JB = cJ , \quad \textit{where } c^2 = (v-1)(\lambda+1) ,$$

(10.1.26) $$B^T = (-1)^{\frac{1}{2}(v-\lambda)+1} B .$$

Proof. The eigenvalues of AA^T are $v-1-\lambda$ with multiplicity $v-1$ and $(v-1)(\lambda+1)$ with multiplicity 1. Since A^TA has the same eigenvalues as AA^T we see that $A^TA-(v-1-\lambda)I$ is a symmetric matrix of rank 1. Therefore

$$A^TA = (v-1-\lambda)I + \alpha\underline{x}\underline{x}^T ,$$

where α is some constant and $\underline{x}^T := (x_1,x_2,\dots,x_v)$. All diagonal elements of A^TA are $v-1$. Therefore, if we take $\alpha = \lambda$, all the x_i are $+1$ or -1 $(i = 1,2,\dots,v)$. Now take $D := \mathrm{diag}[x_1,x_2,\dots,x_v]$. Then $B := AD$ satisfies (10.1.24). If we multiply (10.1.24) by B and $\underline{j}$ we obtain

$$BB^T(B\underline{j}) = (v-1)(\lambda+1)B\underline{j}$$

and since $\underline{j}$ is the only eigenvector of BB^T belonging to the eigenvalue $(v-1)(\lambda+1)$ we find $B\underline{j} = c\underline{j}$. In the same way $B^T\underline{j} = c\underline{j}$. Therefore $c^2 = (v-1)(\lambda+1)$. Now consider row i and row j of B $(i \neq j)$. Let n_0, n_1, n_2 and n_3 denote the number of indices k for which $(b_{ik},b_{jk}) = (1,1)$, $(1,-1)$, $(-1,1)$ and $(-1,-1)$, respectively. Then from $n_0+n_1+n_2+n_3 = v-2$ and the three equations implied by (10.1.24) and (10.1.25) we find

$$(10.1.27) \qquad 4n_1 = v-\lambda-2-(b_{ij}-b_{ji}) .$$

Since $b_{ij}-b_{ji} = 0, +2$, or -2 we find from (10.1.27) that $v-\lambda$ is even and $b_{ij}-b_{ji} \equiv$ $\equiv v-\lambda-2 \pmod 4$, which is (10.1.26).
We remark that A clearly determines B up to multiplication by $-I$.

We now consider some special cases.

(a) Let $\lambda = -1$ in (10.1.24). Then B satisfies (10.1.22), i.e. B is an S-matrix. From Theorem 10.1.4 it follows that S is symmetric or skew.

(b) Let $\lambda = 0$ in (10.1.23), i.e. A is a C-matrix. We have already remarked that we can multiply suitable rows and columns of C to obtain

$$\begin{bmatrix} 0 & \underline{j}^T \\ \underline{j} & S \end{bmatrix} ,$$

where S is an S-matrix. By (a) S is symmetric or skew.

(c) Let $v \equiv \lambda \pmod 4$. By (10.1.26) B is skew. Then (10.1.25) implies $c = 0$, i.e. $\lambda = -1$ (case (a)).

(d) Let $v \equiv \lambda+2 \pmod 4$. Now B is symmetric. Define the symmetric (0,1)-matrix X by

$$2X := J - I - B .$$

If we define $Y := X + I$ then substitution in (10.1.24) and (10.1.25) yields

$$XY = YX = B(v,k,\lambda) , \quad XJ = kJ ,$$

a special case of (10.1.4) which was treated by Bridges and Ryser in [1]. For further references and connections to other parts of combinatorial theory (e.g. tournaments) we refer to [5].

10.2. *Block designs with repeated blocks.*

Suppose that the block design $BD(v,k\,;b,r,\lambda)$ contains exactly t distinct blocks. Let A be the incidence matrix of size t by v formed from the distinct rows of the incidence matrix A' of the block design. Let row i of A occur e_i times in A'. Define $E = \mathrm{diag}[e_1,\dots,e_t]$. Then A satisfies the matrix equations

$$(10.2.1)\qquad A^T E A = (r-\lambda)I + \lambda J\,,$$

$$(10.2.2)\qquad A(1,1,\dots,1)^T = k(1,1,\dots,1)^T\,.$$

We now apply Theorem 10.1.2. We first define a real matrix A^* of order t,

$$(10.2.3)\qquad A^* = [E^{\frac{1}{2}}A, Z]\,,$$

where we require the columns of Z to be orthogonal to the columns of $E^{\frac{1}{2}}A$ and we require Z to be of rank $t-v$. Furthermore we require that

$$(10.2.4)\qquad Z^T Z = (r-\lambda)I\,.$$

It is easily seen that such a matrix Z exists. We then have

$$(10.2.5)\qquad A^{*T}A^* = ((r-\lambda)I + \lambda J) \oplus (r-\lambda)I\,,$$

where the first matrix in the direct sum is of order v and the second matrix is of order $t-v$. In Theorem 10.1.2 we take $X = A^{*T}$, $Y = A^*$, $n = t$, $\lambda_1 = \dots = \lambda_v = \lambda$, $\lambda_{v+1} = \dots = \lambda_t = 0$ and $D = (r-\lambda)I$. Then from (10.1.15), (10.1.16), (10.1.17) we have

$$w = (r-\lambda+\lambda v)(r-\lambda)^{-1}\,,\quad wx_i = wy_i = k\sqrt{\lambda e_i}\,(r-\lambda)^{-1}\,.$$

It follows that

$$(10.2.6)\qquad AA^T = (r-\lambda)E^{-1} + \lambda k r^{-1} J - W\,,$$

where

$$(10.2.7)\qquad W := (E^{-\frac{1}{2}}Z)(E^{-\frac{1}{2}}Z)^T$$

is a symmetric positive semidefinite matrix of order t and rank $t-v$. In (10.2.6) we consider the diagonal of W. We see that $-k + (r-\lambda)e_i^{-1} + \lambda k r^{-1}$ is nonnegative. We thus arrive at the following generalization of Fisher's inequality (H.(10.2.3)) (cf. [6], [7]).

THEOREM 10.2.1. *If in a block design* BD(v,k ; b,r,λ) *a block is repeated* e *times then*

$$(10.2.8) \qquad b/v \geq e .$$

Proof. From $-k + (r-\lambda)e^{-1} + \lambda k r^{-1}$ and the relations bk = vr (H.10.1.1a) and $r(k-1) = \lambda(v-1)$ (H.10.1.1b) the result follows.

Let λ_{ij} denote the inner product of rows i and j ($i \neq j$) of A, i.e. the number of objects that the blocks i and j have in common. The principal submatrix of order 2 of W in (10.2.7) determined by rows i and j of W has a nonnegative determinant. From (10.2.6) we find an expression for this determinant. The result is

THEOREM 10.2.2. *If in a block design* BD(v,k ; b,r,λ) *block* i *occurs* e_i *times and block* j *occurs* e_j *times and these blocks have* λ_{ij} *objects in common, then*

$$(10.2.9) \qquad \left(\frac{r}{e_i} - k\right)\left(\frac{r}{e_j} - k\right) \geq \left(\frac{\lambda k - r\lambda_{ij}}{r - \lambda}\right)^2 , \qquad (i \neq j) .$$

This theorem generalizes a result of W.S. Connor (cf. [8]).
Now suppose for some i we have $e_i = r/k = b/v$. Then (10.2.9) implies $\lambda_{ij} = \lambda k/r = \lambda/e_i$ and therefore $e_i \mid (b,r,\lambda)$. We see that equality in (10.2.8) is not possible if $(b,r,\lambda) = 1$ and there is a repeated block.
We are interested in block designs which have at least one repeated block and $(b,r,\lambda) = 1$. We call these *primitive repetition designs* and use the abbreviation PRD.
If, in a block design, we replace all blocks by their complements we obtain another block design. So we assume $k \leq \frac{1}{2}v$. Given v and k we find for b and r the equations:

$$b = \frac{v(v-1)}{k(k-1)}\lambda , \qquad r = \frac{v-1}{k-1}\lambda$$

where λ must be chosen in such a way that b and r are integers and $(b,r,\lambda) = 1$. If this value of λ is 1 then the design does not have repeated blocks. By (10.2.8) we must have $b > 2v$ if there is to be a repeated block. If we order the possible sets of parameters by increasing v, and for fixed v by increasing k, we find a list starting as follows:

No.	v	k	b , r , λ
1	8	3	56 , 21 , 6
2	10	3	30 , 9 , 2
3	11	3	55 , 15 , 3
4		4	55 , 20 , 6
5	12	3	44 , 11 , 2
6		4	33 , 11 , 3
7		5	132 , 55 , 20
8	13	5	39 , 15 , 5

No.	v	k	b , r , λ
9	14	3	182 , 39 , 6
10		4	91 , 26 , 6
11		5	182 , 65 , 20
12		6	91 , 39 , 15
13	15	4	105 , 28 , 6
14		6	35 , 14 , 5
15	16	3	80 , 15 , 2
16		5	48 , 15 , 4

A block design with at least one repeated block and parameters which have number i in this list will be denoted by PRD_i.
In [9] several constructions for PRD's are described. For $i \le 15$ at least one PRD_i is found but no PRD_{16}. Recently, B. Bettonvil and H. Jonkers found two inequivalent PRD_{16}'s by a computer search. We give one example here of a PRD_1. Let B be the incidence matrix of the design with as blocks all 2-subsets of a set of 7 elements. Then $B^TB = 5I + J$. Let C be the block design consisting of 5 copies of PG(2.2) (cf. H. § 12.2). Then $C^TC = 5(2I + J)$. Now define a BD(8,3 ; 56,21,6) by the incidence matrix

(10.2.10) $$A := \left[\begin{array}{c|c} 1 & \\ 1 & \\ \vdots & B \\ 1 & \\ \hline 0 & \\ 0 & \\ \vdots & C \\ 0 & \end{array}\right].$$

Since

$$A^TA = \left[\begin{array}{c|cccc} 21 & 6 & 6 & \dots & 6 \\ \hline 6 & & & & \\ 6 & & & & \\ \vdots & & B^TB + C^TC & & \\ 6 & & & & \end{array}\right] = 15I + 6J$$

we see that A is indeed the incidence matrix of a block design with the required parameters, i.e. a PRD_1 (since there are 7 blocks which occur 5 times each).

References:

1 W.G. Bridges and H.J. Ryser, Combinatorial Designs and Related Systems, J. of Algebra 13 (1969), 432-446.

2 H.J. Ryser, Symmetric Designs and Related Configurations, J. Comb. Theory (A) 12 (1972), 98-111.

3 H.J. Ryser, A Generalization of the Matrix Equation $A^2 = J$, Linear Algebra and its Applications 3 (1970), 451-460.

4 D.E. Knuth, Notes on Central Groupoids, J. Comb. Theory 8 (1970), 376-390.

5 P. Delsarte, J.M. Goethals and J.J. Seidel, Orthogonal Matrices with Zero Diagonal II, Can. J. Math. 23 (1971), 816-832.

6 H.B. Mann, A Note on Balanced Incomplete Block Designs, Ann. Math. Stat. 40 (1969), 679-680.

7 J.H. van Lint and H.J. Ryser, Block Designs with Repeated Blocks, Discrete Math. 3 (1972), 381-396.

8 W.S. Connor, Jr., On the Structure of Balanced Incomplete Block Designs, Ann. Math. Stat. 23 (1952), 57-71.

9 J.H. van Lint, Block Designs with Repeated Blocks and $(b,r,\lambda) = 1$, J. Comb. Theory (to appear).

XI. DIFFERENCE SETS

After the very extensive treatment in H. Chapter 11 we have only a few remarks. It is worth mentioning that an excellent survey of cyclic difference sets by L.D. Baumert appeared recently (cf. [1]). In the introduction in H. Chapter 11 the idea of a difference set is derived by considering a cyclic block design. From this it follows that the relation (H.10.1.1.b) must hold. Of course this also follows directly from the definition. If $D := \{a_1,a_2,\dots,a_k\}$ is a difference set mod v with every difference $\neq 0$ occurring λ times then obviously

(11.1.1) $$k(k-1) = \lambda(v-1) .$$

If we introduce the parameter $n := k-\lambda$ then from (11.1.1) we find

$$v-1 = 1 + \frac{1}{\lambda}(n+\lambda)(n+\lambda-1) = \lambda + 2n - 1 + \frac{1}{\lambda}n(n-1)$$

and therefore

(11.1.2) $$4n-1 \le v \le n^2+n+1 .$$

If equality holds on the left-hand side of (11.1.2), then we have from (11.1.1), assuming $k \le \frac{1}{2}v$,

(11.1.3) $$(v,k,\lambda) = (4t-1,2t-1,t-1) .$$

These are the parameters of a Hadamard type difference set as e.g. types Q, H_6 and T in H. § 11.6.
If, in (11.1.2), equality holds on the right-hand side, then from (11.1.1) we find $(v,k,\lambda) = (n^2+n+1,n+1,1)$. Hence the cyclic block design corresponding to such a difference set is the point-line incidence matrix of a projective plane as defined in H. § 12.3. The Singer difference sets given by H. Theorem 11.3.1 with n = 3 are examples. Therefore the difference sets S, Q, H_6 and T described in H. § 11.6 are extremal cases of inequality (11.1.2).
It is easily seen that there is a connection between cyclic difference sets and the pseudo-noise sequences discussed in section 9.3. Let $D := \{a_1,a_2,\dots,a_k\}$ be a (v,k,λ)-difference set. We define a (circular) sequence of +1's and -1's by

$$x_i := +1 \quad \text{iff} \quad i \in D \quad (i = 0,1,\dots,v-1) ,$$

$$x_{i+v} = x_i .$$

If $j \neq 0$, then the fact that D is a difference set implies that

(11.1.4) $$\sum_{i=0}^{v-1} x_i x_{i+j} = \lambda + (v-2k+\lambda) - 2(k-\lambda) = v - 4n .$$

This is property (iii) in the definition of a pseudo-noise sequence. Conversely, we can produce difference sets using the methods of section 9.2. If, for instance, we consider the shortened De Bruijn sequence of length 15 produced by the shift register of figure 30, i.e. 000100110101111, we can make a difference set by interpreting the sequence as the characteristic function of this difference set. Reading from right to left we would get $D := \{0,1,2,3,5,7,8,11\}$, a (15,8,4)-difference set (the complement of a Hadamard type set). The construction described in (9.3.1) is not restricted to De Bruijn sequences but it applies to all difference sets with the correct parameters (cf. H.p.207). The particular difference set which we obtained above leads to a Hadamard matrix of order 16. The example (H.14.1.15) shows that such a Hadamard matrix is also obtainable from the (16,6,2) group difference set, first mentioned on H.p.122. We shall give a different description of the group and a short proof that we indeed have a group difference set.

Consider 4-dimensional vector space over GF(2). This is an abelian group of order 16. Let $e_1 := (1,0,0,0)$, $e_2 := (0,1,0,0)$, $e_3 := (0,0,1,0)$, $e_4 := (0,0,0,1)$, $e_5 := (1,1,0,0)$, $e_6 := (0,0,1,1)$. Since $e_1 + e_2 + \ldots + e_6 = (0,0,0,0)$ there is no subset of 4 of the e_i's with sum $(0,0,0,0)$. Hence, if (i_1,j_1), (i_2,j_2) are two distinct pairs with $e_{i_1} + e_{j_1} = e_{i_2} + e_{j_2}$, then we must have $i_1 = j_2$ and $i_2 = j_1$. This proves that among the 30 differences $e_i - e_j$ $(i \neq j;\ i,j = 1,2,\ldots,6)$ each nonzero vector occurs twice, i.e. $\{e_1,e_2,\ldots,e_6\}$ is a (16,6,2)-group difference set. If we interpret the vector space as the additive group of $GF(2^4)$, generated by $x^4 + x + 1$, then the difference set is $\{1,x,x^2,x^3,x^4,x^6\}$. If $a \neq 0$ is an element of $GF(2^4)$ then multiplication of the elements of D by a yields another group difference set. The same is true if we add $b \neq 0$ to all elements of D. The two operations never yield the same difference set. This means that there is no element a which is a multiplier of D in the sense of cyclic difference sets (cf. H.p.132). It has been shown (cf. e.g. [2]) that there is no cyclic (16,6,2) difference set.

References:

[1] L.D. Baumert, Cyclic Difference Sets, Lecture Notes in Mathematics 182, Springer Verlag 1971.

[2] R. Turyn, Character Sums and Difference Sets, Pacific J. Math. 15 (1965), 319-346.

XII. FINITE GEOMETRIES

12.1. *The Theorem of Desargues.*

In the introduction on foundations of finite geometries in H. § 12.1 several theorems are stated without proof. One of these, the first part of H. Theorem 12.1.5, is so easy to prove, using incidences only, that we present its proof here. In this way we also have an opportunity to correct figure 12.1 on H.p.168 which is misleading because it suggests that A_1, A_2 and A_3 are collinear.

THEOREM 12.1.1. *If, in a projective geometry the theorem of Pappus holds, then the theorem of Desargues also holds.*

Proof. Using the notation of H. Theorem 12.1.2 (Pappus) we denote the assertion of that theorem, namely that A_3, B_3 and C_3 are collinear by

$$\text{Pappus}\begin{pmatrix} A_1 & B_1 & C_1 \\ A_2 & B_2 & C_2 \end{pmatrix} = \ell(A_3,B_3,C_3) .$$

Now consider figure 31 (which is an extended version of H.fig.12.1). We define 4 points S, T, U, V by:

A_1B_1 intersects B_2C_2 in S,
OS intersects A_1C_1 in T,
OB_1 intersects A_1C_2 in U,
OS intersects A_2C_2 in V.

Now we have the following applications of the theorem of Pappus:

$$\text{Pappus}\begin{pmatrix} O & C_1 & C_2 \\ A_1 & S & B_1 \end{pmatrix} = \ell(A_3,U,T) , \quad \text{i.e. UT intersects } SC_2 \text{ in } A_3 ,$$

$$\text{Pappus}\begin{pmatrix} O & A_1 & A_2 \\ C_2 & B_2 & S \end{pmatrix} = \ell(C_3,V,U) , \quad \text{i.e. } A_1S \text{ intersects UV in } C_3 ,$$

and therefore

$$\text{Pappus}\begin{pmatrix} A_1 & C_2 & U \\ V & T & S \end{pmatrix} = \ell(A_3,C_3,B_3) ,$$

which is the assertion of the theorem of Desargues.

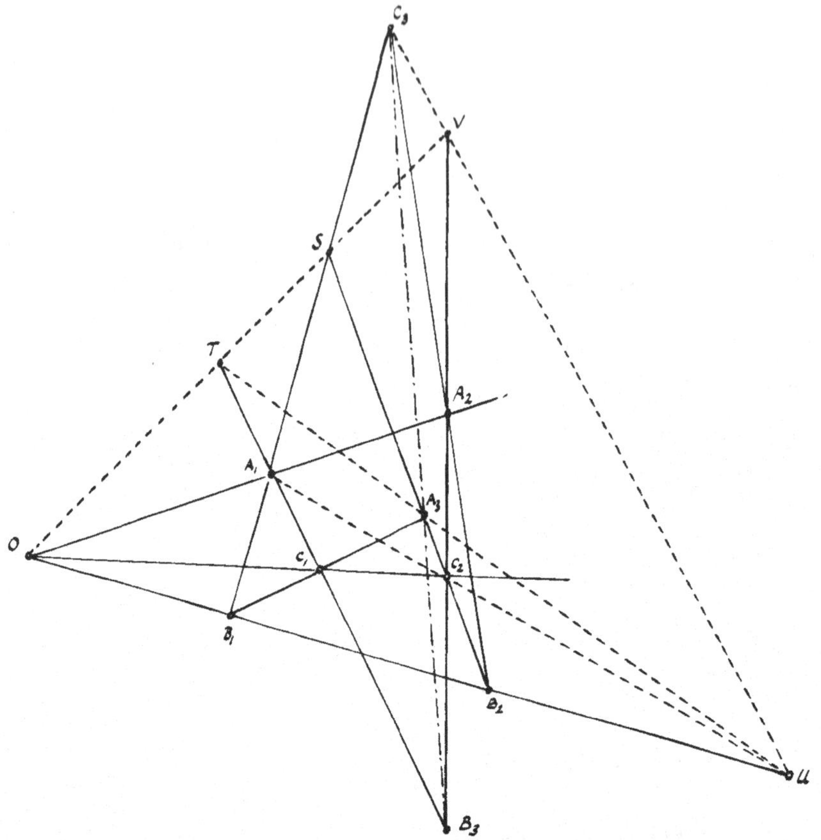

Fig. 31

In H. § 12.4 examples are given of projective planes for which the theorem of Desargues does not hold. If we allow the plane to be infinite, a simple example can be given by constructing what is called the free projective plane generated by 4 points. This is a projective plane in which no "accidental" incidences occur. What we mean will become clear in the following construction.

We start with a set of 4 points, $P_0 := \{p_1, p_2, p_3, p_4\}$. We consider this to be a geometry with no lines. Next we let the geometry P_1 consist of the 4 points of P_0 and the 6 lines $\{p_i, p_j\}$ $(i \neq j)$. In P_1 every 2 points determine a line, (*). We define P_2 by adding a new point for every two lines of P_1 which do not have a common point in P_1. Then P_2 has 7 points $p_1, p_2, \ldots, p_7$ and the lines $\{p_1, p_2, p_5\}$, $\{p_1, p_3, p_7\}$, $\{p_2, p_4, p_6\}$, $\{p_2, p_3, p_6\}$, $\{p_2, p_4, p_7\}$, $\{p_3, p_4, p_5\}$. In P_2 every two lines have a common point, (**). We continue in this way, alternately adjoining new points or new lines and forming a series of geometries P_n which have property (*) if n is odd and property (**) if n is even. Furthermore, the points (lines) added at step n are necessary to assure the property (**) (resp. (*)). Now the free plane P generated by P_0 is defined to be the set of points of $\bigcup_{i=0}^{\infty} P_i$ with as lines the subsets ℓ of P for which $\ell \cap P_n$ is a line in P_n for all sufficiently large n. P is clearly a projective geometry. Suppose P contains a set S of 10 points and 10 lines which form a Desargues configuration. Then there is a minimal index n such that $S \subset P_n$. Every point of S is on 3 lines and every line of S contains 3 points. Let n be even (n odd is treated analogously). Then S contains a point p which is not in P_{n-1}, i.e. p is on only 2 lines of P_n and hence not on 3 lines of S, a contradiction. Therefore P does not contain a Desargues configuration on 10 distinct points. It is easily seen that the points 0, A_1, etc. of H. Theorem 12.1.1 can be chosen in P in such a way that they are distinct. Hence P is non-Desarguesian plane.

12.2. *Automorphisms.*

In H. § 12.2 it is noted that the full collineation group of a projective geometry of dimension n over GF(q) is the product of the group of nonsingular matrices of order n + 1 over GF(q) and the group of automorphisms of the field. We consider this question for the more general case of a Desarguesian projective plane P coordinated by a field F as in H. Theorem 12.1.4. To see that we have the same type of automorphism group it is sufficient to consider an automorphism φ of P which fixes 4 points not on a line. Without loss of generality we may take these points to be represented by the 3-tuples (1,0,0), (0,1,0), (0,0,1), (1,1,1). The 3-tuples (x_1,x_2,x_3) with $x_3 = 0$ form a line in P which is fixed (as a line) by φ since two of its points are fixed. The 3-tuples (x_1,x_2,x_3) with $x_3 \neq 0$ form an *affine* plane A. In A we represent points by inhomogeneous coordinates (ξ_1,ξ_2) where $\xi_i := x_i/x_3$ (i = 1,2). Lines of P correspond to lines of A. We say two lines of A are parallel if the corresponding lines of P intersect on the line $x_3 = 0$. Since φ fixes this line we see that φ is an automorphism of A which maps parallel lines into parallel lines. We define the mapping $\sigma: F \to F$ by $(\sigma(a),0) := \varphi(a,0)$. Clearly the lines given by $\xi_1 = a$, resp. $\xi_1 = \sigma(a)$ are parallel and therefore φ maps the first of these into the second. Since the line $\xi_1 = \xi_2$ is fixed in A (two fixed points) φ maps (a,a) into $(\sigma(a),\sigma(a))$. From this it follows that φ maps the line $\xi_2 = a$ into the line $\xi_2 = \sigma(a)$. Finally, it follows that $\varphi(a,b) = (\sigma(a),\sigma(b))$. Note that σ is a one to one mapping of F onto F. Now consider figure 32a.

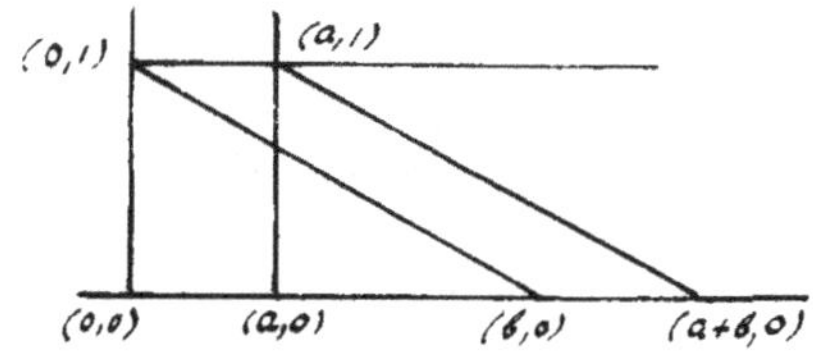

Fig. 32a

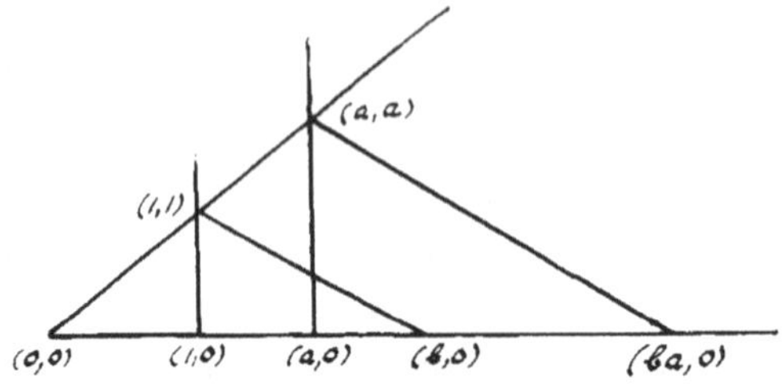

Fig. 32b

The lines through (0,1) and (b,0), respectively (a,1) and (a+b,0), are parallel (they correspond to the lines $x_1 + bx_2 - bx_3 = 0$ and $x_1 + bx_2 - (a+b)x_3 = 0$ in P which intersect in (b,-1,0)). Now consider the action of φ. We have $\varphi(0,1) = (0,1)$, $\varphi(a,0) = (\sigma(a),0)$, $\varphi(b.0) = (\sigma(b),0)$. Since φ maps parallel lines into parallel lines we find from figure 32a that φ maps (a+b,0) into $(\sigma(a)+\sigma(b),0)$. We have thus proved that $\sigma(a+b) = \sigma(a) + \sigma(b)$ for all $a,b \in F$. In the same way we have from figure 32b, where the lines through (b,0) and (1,1), respectively (ba,0) and (a,a) are parallel, that $\sigma(ba) = \sigma(b)\sigma(a)$ for all $a,b \in F$. Therefore we have proved that σ is an automorphism of the field F. The action of φ on the plane P is described by $\varphi(x_1,x_2,x_3) = (\sigma(x_1),\sigma(x_2),\sigma(x_3))$.

12.3. *Near-fields.*

In H. § 12.4 near-fields are introduced and then described without proof that the definition is satisfied. It seems worthwhile to give a few more details here.

DEFINITION. *Consider a set* K *with two binary operations* +, $\circ$. *We shall call* (K,+,$\circ$) *a near-field if* (K,+) *is an abelian group,* (K,$\circ$) *is a group, and*

$$\forall_{a\in K}\,\forall_{b\in K}\,\forall_{m\in K}\,[(a+b)\circ m = (a\circ m) + (b\circ m)]\ .$$

Let $q = p^h$ be a power of a prime p and let v be an integer, all of whose prime factors divide $q-1$. Furthermore, let $v \not\equiv 0 \pmod 4$ if $q \equiv 3 \pmod 4$. Let $r := hv$, $n := p^r = q^v$. We first show that the multiplicative order of q (mod $v(q-1)$) is v. Let π be any prime which divides $q-1$, say $\pi^\sigma \,\|\, (q-1)$. Let $\pi^\tau \,\|\, v$. Let $m = m_1\pi^t$ where $(m_1,\pi) = 1$. Then

$$q^m - 1 = (q^{\pi^t} - 1)(1 + q^{\pi^t} + \ldots + (q^{\pi^t})^{m_1-1})\ ,$$

and since the second factor is congruent to $m_1 \not\equiv 0 \pmod \pi$ we see that the highest power of π dividing $q^m - 1$ divides q^{π^t}. Now

$$(*)\qquad q^{\pi^t} - 1 = (q^{\pi^{t-1}} - 1)(1 + q^{\pi^{t-1}} + \dots + (q^{\pi^{t-1}})^{\pi-1}) .$$

We define $e(t)$ by $\pi^{e(t)} \| (q^{\pi^t} - 1)$. Let $q^{\pi^{t-1}} = 1 + \ell\pi^{e(t-1)}$ with $(\ell,\pi) = 1$. Then from $(*)$ we have

$$q^{\pi^t} - 1 = \ell\pi^{e(t-1)}\{\pi + \ell\pi^{e(t-1)} \frac{\pi(\pi-1)}{2} + \dots\}$$

from which it follows that

$$e(t) = e(t-1) + 1$$

unless $e(t-1) = 1$ and $\pi = 2$. Hence, with the exception $q \equiv 3 \pmod 4$ and $\pi = 2$, we find that $\pi^{\sigma+t} \| (q^m - 1)$ if $\pi^t \| m$. Therefore the multiplicative order of q (mod $v(q-1)$) is indeed v, because in the exceptional case $q \equiv 3 \pmod 4$ and $\pi = 2$ we have required that $\tau \le 1$. From this fact it follows that a 1-1 mapping α of the integers $j \in [0,v-1]$ is defined by

$$(12.3.1)\qquad q^{\alpha(j)} \equiv 1 + j(q-1) \pmod{v(q-1)} ,$$

where $0 \le \alpha(j) \le v-1$.

Let $K := GF(q^v)$ and let z be a fixed primitive root. Let $+$ be the addition in $GF(q^v)$. If $u \in K$, $u \neq 0$, then $u = z^{kv+j}$ with $0 \le j \le k-1$. We define

$$(12.3.2)\qquad \beta(u) := q^{\alpha(j)} ,$$

$$(12.3.3)\qquad w \circ u := w^{\beta(u)}u \quad \text{(multiplication in } GF(q^v)) .$$

(If $u = 0$ we define $w \circ u = u \circ w := 0$.) We claim that $(K,+,\circ)$ is a near-field. We start by showing that multiplication is associative. Let $b = z^{k_1v+j_1}$, $c = z^{k_2v+j_2}$. Then by (12.3.3) we have

$$(12.3.4)\qquad (a \circ b) \circ c = (a^{\beta(b)}b)^{\beta(c)}c = a^{\beta(b)\beta(c)}b^{\beta(c)}c .$$

Furthermore, by (12.3.1), (12.3.2), (12.3.3)

$$(12.3.5)\qquad b \circ c = b^{\beta(c)}c = z^{(k_1v+j_1)q^{\alpha(j_2)}+k_2v+j_2} =$$

$$= z^{(k_1v+j_1)(1+j_2(q-1)+mv(q-1))+k_2v+j_2} =$$

$$= z^{k_3v+j_1+j_2+j_1j_2(q-1)} .$$

Since

$$\beta(b)\beta(c) = q^{\alpha(j_1)+\alpha(j_2)} \equiv \{1 + j_1(q-1)\}\{1 + j_2(q-1)\} \pmod{v(q-1)}$$

$$\equiv 1 + \{j_1 + j_2 + j_1 j_2(q-1)\}(q-1) \pmod{v(q-1)}$$

we find from (12.3.5) that $\beta(b \circ c) = \beta(b)\beta(c)$. Hence

$$a \circ (b \circ c) = a^{\beta(b)\beta(c)}(b \circ c) = a^{\beta(b)\beta(c)} b^{\beta(c)} c \, . \tag{12.3.6}$$

From (12.3.4) and (12.3.6) the associativity follows.

Suppose in (12.3.3) $w \circ u = w$, $w = z^a$, $u = z^{kv+j}$. Then from (12.3.1), (12.3.2) we find

$$a(q^{\alpha(j)} - 1) + kv + j \equiv 0 \pmod{q^v - 1} \, ,$$

$$j\{1 + a(q-1)\} + kv \equiv 0 \pmod{v(q-1)}$$

and from the condition on v and $q-1$ it then follows that $j = 0$ and therefore $u = 1$.

If $u \in GF(q^v)$, $u \neq 0$ and u^{-1} is the inverse in $GF(q^v)$, then u^* is uniquely defined by $(u^*)^{\beta(u)} = u^{-1}$. Then we have

$$u^* \circ u = (u^*)^{\beta(u)} u = 1$$

and

$$u^* \circ (u \circ u^*) = (u^* \circ u) \circ u^* = 1 \circ u^* = u^*$$

which implies that $u \circ u^* = 1$. Therefore, every nonzero element of K has a (unique) inverse in K with respect to the operation $\circ$.

It remains to show that one distribution law holds. We have, because $\beta(c)$ is a power of q,

$$(a+b) \circ c = (a+b)^{\beta(c)} c = (a^{\beta(c)} + b^{\beta(c)})c = (a \circ c) + (b \circ c) \, .$$

XIII. ORTHOGONAL LATIN SQUARES

In this chapter we shall survey the presently known lower bounds for N(n), i.e. the number of mutually orthogonal latin squares of order n, for $n \leq 100$. We use the convention $N(0) := N(1) := \infty$. We shall use the symbol

$$n_r := \min \{k \mid N(n) \geq r \text{ for all } n \geq k\} .$$

If, in a pairwise balanced design $BIB(v,k_1,\dots,k_m,\lambda)$ (cf. H.p.196), the first r equiblock components form a clear set we shall denote the design by $BIB(v,k_1,\dots,k_r; k_{r+1},\dots,k_m,\lambda)$. Besides orthogonal arrays, as defined on H.p.190, we use the concept of *transversal system* $T_0(m,t)$ as defined on H.p.224. From the definitions we see that the existence of a $T_0(m,t)$, the existence of an OA(t,m), and the assertion $N(t) \geq m-2$ are equivalent.

13.1. *Applications of H. Theorem 13.3.2.*

The best known lower bound for N(82) is due to H. Hanani (cf. [1]). Consider the plane EG(2,11) embedded in PG(2,11). Let the points of EG(2,11) be coordinated in the usual way as (x,y), $x = 0,1,\dots,10$; $y = 0,1,\dots,10$. Now leave out the points (x,y) with $0 \leq x \leq 3$, $1 \leq y \leq 10$ and add one point of the line at infinity in PG(2,11) which is not on the line $x = 0$ or the line $y = 0$. With these points only, we consider the lines of PG(2,11) as blocks. It is easily seen that each line contains 7, 8, 9, or 11 points of the 82 points under consideration. Therefore, we have constructed a BIB(82,7,8,9,11,1). By H. Theorem 13.3.2 we have

(13.1.1) $\qquad N(82) \geq 5$.

Next we consider the projective plane PG(2,8). We coordinate by the elements of GF(8) (cf. H.Ch.12). Let C be the set of 10 points consisting of (1,0,0), (0,1,0), and $(\xi,\xi^2,1)$ where ξ runs through GF(8). Clearly no 3 of these points are collinear. C is usually called an *oval* with *nucleus* (cf. [2], § 12.3).
Let $1 \leq x \leq 10$. From PG(2,8) we delete a subset of x points of C. Clearly each line of PG(2,8) contains 7, 8 or 9 of the remaining points. Hence we have constructed a BIB(73-x,7,8,9,1) for each x between 1 and 10. By H. Theorem 12.3.2 we have

(13.1.2) $\qquad N(73-x) \geq 5 \quad$ for $x = 1,2,\dots,10$.

This construction is due to R.C. Bose and S.S. Shrikhande ([3]). They also gave a construction of this type starting with the euclidean plane EG(2,9). In this case one can take for C the set of 10 points of a conic which does not meet the line at infinity (cf. [2], § 2.7). The same construction as above then leads to

(13.1.3) $\qquad N(81-x) \geq 5 \quad$ for $x = 1,2,\dots,10$.

13.2. *Two generalizations of H. Theorem 13.3.2.*

We first remark that the description of the construction of an OA(n,k+1) from $k-1$ mutually orthogonal latin squares of order n (H.p.190) shows that the array has the form

$$(13.2.1)\qquad \left[\begin{array}{c|c|c|c|c} 1\ 1\ \ldots\ 1 & 2\ 2\ \ldots\ 2 & 3\ 3\ \ldots\ 3 & & n\ n\ \ldots\ n \\ 1\ 2\ \ldots\ n & & & & \\ 1\ 2\ \ldots\ n & A_1 & A_2 & & A_{n-1} \\ \vdots\qquad\ \vdots & & & & \\ 1\ 2\ \ldots\ n & & & & \end{array}\right]$$

where each row of A_i is a permutation of the symbols $1,2,\ldots,n$ $(i = 1,2,\ldots,n-1)$.
Next we note that it follows from H. Theorem 5.1.9 that the incidence matrix of a symmetric BIB(v,k,1) is the sum of k permutation matrices. If we denote the points of the design as $1,2,\ldots,v$ and write the blocks as columns of a k by v array, then by the previous remark the columns can be written in such a way that each row of the array is a permutation of the symbols $1,2,\ldots,v$. We call this the standard form.
Now, consider a symmetric BIB(v,k,1) and analyse the proof of H. Theorem 13.3.2. Again, let $c := N(k) + 1$. Suppose the design is in standard form and consider the array $C = (B_1,B_2,\ldots,B_v,E)$ of (H.13.3.9). Reorder the columns of C by first taking the first column of $B_1,B_2,\ldots,B_v$, respectively; then the second column of each, etc. We leave E at the end. The resulting matrix has the form

$$C^* = (C_1,C_2,\ldots,C_{v-1},E)\ ,$$

where each row of C_i resp. E is a permutation of $1,2,\ldots,v$.
If we add the row $(1,1,\ldots,1,2,2,\ldots,2,\ldots,v,v,\ldots,v)$ to C^* the result is clearly an OA(v,c+1). Therefore we have the following improvement of H. Theorem 13.3.2:

THEOREM 13.2.1. *If there is a symmetric* BIB(v,k,1) *then*

$$(13.2.2)\qquad N(v) \geq N(k)\ .$$

From the existence of PG(2,s) if s is a prime power we find the corollary:

COROLLARY. *If* s *is a prime power, then*

$$(13.2.3)\qquad N(s^2+s+1) \geq N(s+1)\ .$$

(The inequalities (13.2.2) and (13.2.3) are also from [3].)
We give one more result using symmetric designs. It concerns a generalization of H. Theorem 13.3.5(3) (see [4]).

THEOREM 13.2.2. *If there is a symmetric* BIB(v,k,1), *then*

$$(13.2.4)\qquad N(k^2+1) \geq \min\{N(k),\ N(k+1)-1\}\ .$$

Proof. Let $c = N(k+1) + 1$. As in the proof of H. Theorem 13.3.2 we consider the c by $k^2 + k$ matrix A obtained from the array $OA(k+1,c+1)$, in the form of (13.2.1), by deleting the first row and the first $k+1$ columns. Let the block design $BIB(v,k,1)$ be on the points $1,2,\dots,v$ and consider k new points $x_1,x_2,\dots,x_k$. We proceed as in the proof of H. Theorem 13.3.2 but the substitutions in different submatrices of A now differ.

First, let BIB_i denote the block design obtained from the symmetric $BIB(v,k,1)$ by adjoining x_i to each block. If $A = (A_1\ A_2\ \dots\ A_k)$ then, for each block of BIB_i we replace the elements $1,2,\dots,k+1$ of the matrix A_i by the elements $1,2,\dots,k,x_i$ of the block. Just as in (H.13.3.9) this results in an array C which in this case has c rows and $vk(k+1)$ columns. If $c' := N(k) + 2$ we consider the $OA(k,c')$ on the symbols $x_1,x_2,\dots,x_k$. Call this C'. Finally, let E be made up of v columns, each column of E consisting of the same number i repeated $\bar{c} := \min\{c,c'\}$ times $(i = 1,2,\dots,v)$. Consider the first $\bar{c}$ rows of C, C' and E. Together these form an $OA(v+k,\bar{c})$ on the symbols $1,2,\dots,v,x_1,x_2,\dots,x_k$. This proves the theorem

Note that the first interesting application of this theorem is the case $k = 3$ where $PG(2,2)$ leads to the result $N(10) \geq 2$. In this case we have

$$(A_1,A_2,A_3) = \begin{bmatrix} 1&2&3&4&1&2&3&4&1&2&3&4\\ 2&1&4&3&3&4&1&2&4&3&2&1\\ 3&4&1&2&4&3&2&1&2&1&4&3\\ 4&3&2&1&2&1&4&3&3&4&1&2 \end{bmatrix},$$

$$BIB_i = \begin{bmatrix} 1&2&3&4&5&6&7\\ 2&3&4&5&6&7&1\\ 4&5&6&7&1&2&3\\ x_i&x_i&x_i&x_i&x_i&x_i&x_i \end{bmatrix}, \quad (i = 1,2,3),$$

$$C' = \begin{bmatrix} x_1&x_1&x_1&x_2&x_2&x_2&x_3&x_3&x_3\\ x_1&x_2&x_3&x_1&x_2&x_3&x_1&x_2&x_3\\ x_1&x_2&x_3&x_2&x_3&x_1&x_3&x_1&x_2\\ x_1&x_2&x_3&x_3&x_1&x_2&x_2&x_3&x_1 \end{bmatrix},$$

$$E = \begin{bmatrix} 1&2&3&4&5&6&7\\ 1&2&3&4&5&6&7\\ 1&2&3&4&5&6&7\\ 1&2&3&4&5&6&7 \end{bmatrix}.$$

Each column of A yields 7 columns of C.

For $k = 8$ the bound which (13.2.4) yields is better than our other theorems can do.

13.3. *Some values of* N(n).

In a recent paper A. Hedayat ([5]) exhibited 3 mutually orthogonal latin squares of order 15.
The result $N(46) \geq 3$ occurs in a paper by C.-C. Shih ([6], Math. Rev. 37, # 5113). This result can be obtained from H. Theorem 13.3.2 (R.M. Wilson, private communication). Since PG(2,8) is a BIB(73,9,1) we find from H. Theorem 13.3.4

$$N(70) \geq 6 \; . \tag{13.3.1}$$

Since EG(2,9) is a BIB(81,9,1) we find from the same theorem

$$N(78) \geq 6 \; . \tag{13.3.2}$$

If we apply the first part of H. Theorem 13.3.3 to EG(2,9) we find

$$N(80) \geq 7 \; . \tag{13.3.3}$$

13.4. *Three recent theorems of R.M. Wilson.*

In a recent paper R.M. Wilson ([7]) proved a number of theorems which generalize results of Hanani ([1]). The proofs are rather complicated but the theorems are so much stronger than previous results that we consider it worthwhile to give all the details. We have adapted the terminology.

THEOREM 13.4.1. *Let* $G_1,G_2,\ldots,G_k,H_1,H_2,\ldots,H_\ell$ *be* $k+\ell$ *disjoint sets of size* t, $G := \cup G_i$, $H := \cup H_j$. *Let there be a transversal system* $\mathcal{T} := T_0(k+\ell,t)$ *on the set* $X := G \cup H$ *(subdivided as above) with blocks* Y_n $(n = 1,2,\ldots,t^2)$. *Let* $S \subset H$, $|S| = s$. *Define* $\eta_n := |Y_n \cap S|$ $(n = 1,2,\ldots,t^2)$, $h_j := |S \cap H_j|$ $(j = 1,2,\ldots,\ell)$. *Let* $m \geq 1$. *If*

(i) *for* $j = 1,2,\ldots,\ell$ *there exists a* $T_0(k,h_j)$,

and

(ii) *for* $n = 1,2,\ldots,t^2$ *there exists a* $T_0(k,m+\eta_n)$ *which has* η_n *mutually disjoint blocks,*

then there exists a $T_0(k,mt+s)$.

Proof. We first remark that the number of pairs of elements of S contained in the blocks Y_n is $\sum_{n=1}^{t^2} \binom{\eta_n}{2}$ on the one hand, but on the other hand by the definition of a transversal system it is equal to $\sum_{i \neq j} h_i h_j$. It follows that

$$2\sum_{n=1}^{t^2} \binom{\eta_n}{2} + \sum_{j=1}^{\ell} h_j^2 = (h_1 + h_2 + \ldots + h_\ell)^2 = s^2 \; . \tag{13.4.1}$$

We use the following notation:

$$Y'_n := Y_n \cap G\ ,\quad Y''_n := Y_n \cap S \quad (\text{i.e. } |Y''_n| = \eta_n)\ ,$$

$$I_k := \{1,2,\dots,k\}\ ,\quad M \text{ is some m-set, say } I_m\ .$$

Let the set $X^* := (G \times M) \cup (I_k \times S)$ of cardinality $k(mt+s)$ be divided into the k disjoint subsets

$$G_i^* := (G_i \times M) \cup (\{i\} \times S) \qquad (i = 1,2,\dots,k)\ .$$

For $n = 1,2,\dots,t^2$ we construct a transversal system $\mathcal{T}_n = T_0(k,m+\eta_n)$ on the set $(Y'_n \times M) \cup (I_k \times Y''_n)$, divided into the k disjoint subsets

$$((Y'_n \cap G_i) \times M) \cup (\{i\} \times Y''_n) \qquad (i = 1,2,\dots,k)$$

and we do this in such a way that $I_k \times \{z\}$ is a block of $\mathcal{T}_n$ for every $z \in Y''_n$. This is possible by condition (ii). We delete the η_n disjoint blocks $I_k \times \{z\}$. The set consisting of the remaining $(m+\eta_n)^2 - \eta_n$ blocks of $\mathcal{T}_n$ is denoted by $\mathcal{T}_n^*$. Define $\mathcal{T}^* := \bigcup_{n=1}^{t^2} \mathcal{T}_n^*$. Next we construct a transversal system $\mathcal{T}_j^{**} = T_0(k,h_j)$ on the set $I_k \times (S \cap H_j)$ which is divided into the k disjoint subsets $\{i\} \times (S \cap H_j)$, $(i = 1,2,\dots,k)$. Let $\mathcal{T}^{**} := \bigcup_{j=1}^{\ell} \mathcal{T}_j^{**}$. It is clear that the sets of $\mathcal{T}^* \cup \mathcal{T}^{**}$ are transversals of $X^* = G_1^* \cup G_2^* \cup \dots \cup G_k^*$. The total number of transversals in this set is

$$\sum_{n=1}^{t^2} \{(m+\eta_n)^2 - \eta_n\} + \sum_{j=1}^{\ell} h_j^2 =$$

$$= m^2t^2 + 2m \sum_{n=1}^{t^2} \eta_n + 2 \sum_{n=1}^{t^2} \binom{\eta_n}{2} + \sum_{j=1}^{\ell} h_j^2 = (mt+s)^2$$

by (13.4.1). To complete the proof that $\mathcal{T}^* \cup \mathcal{T}^{**}$ is a $T_0(k,mt+s)$ it now suffices to remark that:

(a) the blocks of $\mathcal{T}^{**}$ contain all pairs of type $\{(i_1,z_1),(i_2,z_2)\}$ where $i_1 \neq i_2$ and $\{z_1,z_2\} \subset H_j$ for some j,

(b) if $g_1 \in G_i$, $g_2 \in G_j$ $(i \neq j)$, then there is an n such that $\{g_1,g_2\} \subset Y_n$ and hence $\{(g_1,\mu_1),(g_2,\mu_2)\}$ occurs in some block of $\mathcal{T}_n^*$ if $\mu_1 \in M$, $\mu_2 \in M$. Furthermore, the blocks of $\mathcal{T}^*$ contain all pairs $\{(g,\mu),(i,z)\}$ where $g \notin G_i$ and all pairs $\{(i_1,z_1),(i_2,z_2)\}$ where $i_1 \neq i_2$, $z_1 \neq z_2$, $z_1 \in S$, $z_2 \in S$.

The following two theorems are applications.

THEOREM 13.4.2. *If* $0 \le u \le t$, *then*

(13.4.2) $N(mt+u) \geq \min\{N(m),N(m+1),N(t)-1,N(u)\}$.

Proof. Let $k := 2 + \min\{N(m),N(m+1),N(t)-1,N(u)\}$. The transversal systems $T_0(k,m)$, $T_0(k,m+1)$, $T_0(k+1,t)$, $T_0(k,u)$ exist. In Theorem 13.4.1 take $\ell = 1$ and S a subset of H_1 with $|S| = u$ (i.e. $s = u$). Then $h_1 = u$ and $n_n = 0$ or 1 for all blocks Y_n of $T_0(k+1,t)$. The conditions (i) and (ii) are satisfied. Hence a $T_0(k,mt+u)$ exists, which is equivalent to the assertion $N(mt+u) \geq k-2$.

Example. Take $m = u = 4$, $t = 5$. We find

(13.4.3) $N(24) \geq 3$,

which also follows from H. Theorem 13.3.3 applied to EG(2,5).

THEOREM 13.4.3. *If* $0 \leq u \leq t$, $0 \leq v \leq t$, *then*

(13.4.4) $N(mt+u+v) \geq \min\{N(m),N(m+1),N(m+2),N(t)-2,N(u),N(v)\}$.

Proof. Again, let $k-2$ be the indicated minimum. In Theorem 13.4.1 we now take $\ell = 2$ and choose S such that $|S \cap H_1| = u$, $|S \cap H_2| = v$. Condition (i) of the theorem is satisfied. For every block Y_n of $T_0(k+2,t)$ we have $n_n = 0$, 1 or 2. Transversal systems $T_0(k,m+i)$ exist for $i = 0$, 1 and 2. We must show that a $T_0(k,m+2)$ has 2 disjoint blocks. For each point of Y_1 there are $m+1$ other blocks Y_i containing this point. To show that there is a block disjoint from Y_1 it suffices to show that $k(m+1) < (m+2)^2 - 1$. This follows from $k \leq N(m)+2 \leq m+1$. The existence of a $T_0(k,mt+u+v)$ implies (13.4.4).

Example. Take $m = t = 7$, $u = v = 1$. Then from (13.4.4) we find

(13.4.5) $N(51) \geq 4$.

13.5. *Lower bounds for* $N(n)$, $2 \leq n \leq 100$.

In the following table we give the best lower bounds known to us for $N(n)$ and a theorem or reference for this value. We omit prime powers. If the best bound is obtained by factoring n and applying H. Theorem 13.2.2 we just give the factors.

n	lower bound for N(n)	references	n	lower bound for N(n)	references
6	1	Tarry [8]	58	5	58 = 7.8 + 1 + 1 (13.4.4)
10	2	H. Theorem 13.3.1	60	4	60 = 5.12
12	5	H.p.203	62	4	62 = 7.8 + 5 + 1 (13.4.4)
14	2	H.p.201	63	6	63 = 7.9
15	3	[5]	65	7	k = 8 in (13.2.4)
18	2	H.p.198	66	5	x = 7 in (13.1.2)
20	3	20 = 4.5	68	5	x = 5 in (13.1.2)
21	4	s = 4 in (13.2.3)	69	5	x = 4 in (13.1.2)
22	2	H. Theorem 13.3.1	70	6	(13.3.1)
24	3	(13.4.3)	72	7	72 = 8.9
26	2	H.p.201	74	5	x = 7 in (13.1.3)
28	3	28 = 4.7	75	5	x = 6 in (13.1.3)
30	2	30 = 3.10	76	5	x = 5 in (13.1.3)
33	3	33 = 4.7 + 5 (13.4.2)	77	6	77 = 7.11
34	2	H. Theorem 13.3.1	78	6	(13.3.2)
35	4	35 = 5.7	80	7	(13.3.3)
36	3	36 = 4.9	82	5	(13.1.1)
38	2	H.p.202	84	6	84 = 7.11 + 7 (13.4.2)
39	3	39 = 4.8 + 7 (13.4.2)	85	6	85 = 7.11 + 8 (13.4.2)
40	4	40 = 5.8	86	6	86 = 7.11 + 9 (13.4.2)
42	2	42 = 3.14	87	6	87 = 7.11 + 9 + 1 (13.4.4)
44	3	44 = 4.11	88	7	88 = 8.11
45	4	45 = 5.9	90	5	90 = 11.8 + 1 + 1 (13.4.4)
46	3	[6]	91	6	91 = 7.13
48	3	48 = 4.12	92	6	92 = 7.13 + 1 (13.4.2)
50	5	H.p.199	93	6	93 = 7.11 + 8 + 8 (13.4.4)
51	4	(13.4.5)	94	6	94 = 7.11 + 8 + 9 (13.4.4)
52	3	52 = 4.13	95	6	95 = 8.11 + 7 (13.4.2)
54	4	H.p.199	96	7	96 = 8.11 + 8 (13.4.2)
55	4	55 = 5.11	98	6	98 = 7.13 + 7 (13.4.2)
56	6	56 = 7.8	99	8	99 = 9.11
57	7	s = 7 in (13.2.3)	100	6	100 = 7.13 + 9 (13.4.2)

13.6. *The function* n_r.

Clearly $n_2 = 6$. Theorems like H. Theorem 13.3.8 and the theorems of § 13.4 make it possible to give upper bounds for n_r. Suppose, for example, that we show by the methods of this chapter that $N(n) \geq 3$ for $47 \leq n \leq 332$. In Theorem 13.4.3 we can take m from this interval, $t = 7$, and u and v both from the set $\{0,1,4,5,7\}$. It then follows that $N(n) \geq 3$ for $333 \leq n \leq 2326$ and proceeding in this fashion it follows by induction that $N(n) \geq 3$ for $n \geq 47$.

By using such methods the following bounds have been obtained:

(a) $n_3 \leq 46$ (Wilson [7]). This can be lowered to $n_3 \leq 42$ by reference [6].

(b) $n_4 \leq 52$ (see list in § 13.5).

(c) $n_5 \leq 62$ (Hanani [1]).

(d) $n_6 \leq 90$ (Wilson [7]).

(e) $n_{29} \leq 34115553$ (Hanani [1]).

Result (d) is very good because of the lucky circumstance that 7, 8 and 9 are prime powers. To find an upper bound for n_7 we first apply Theorem 13.4.1 once more.

Let $m \geq 15$ and let $N(m+3) \geq 7$. We consider a $T_0(9,m+3)$. We have already seen that there are two disjoint blocks Y_1, Y_2. For $i = 1,2$ there are $9(m+2)$ blocks which intersect Y_i in a point. Hence there are $(m+3)^2 - 1 - 18(m+2) > 0$ blocks which do not intersect Y_1 or Y_2. This allows us to apply Theorem 13.4.1 with $k = 9$, $\ell = 3$. In exactly the same way as in Theorem 13.4.3 we find:

THEOREM 13.6.1. *If* $0 \leq u_i \leq t$ *for* $i = 1,2,3$; *if* $N(m+i) \geq 7$ *for* $i = 0,1,2,3$ *and* $N(u_i) \geq 7$ *for* $i = 1,2,3$ *and if* $N(t) \geq 10$, *then* $N(mt + u_1 + u_2 + u_3) \geq 7$.

Corollary. If $N(n) \geq 7$ for $a \leq n \leq b$ then $N(n) \geq 7$ for $11a + 16 \leq n \leq 11b - 2$.

Proof. We can apply Theorem 13.6.1 with $a \leq m \leq b - 3$, $u_i = 0, 1, 8, 9$ or 11 $(i = 1,2,3)$ and $t = 11$. Since any integer between 16 and 31 is the sum of three u_i's the result follows.

A computer program was written by F.C. Bussemaker and H.J.L. Kamps which produced bounds for $N(n)$. To start with, a number of known estimates e.g. the table in § 13.5, results from [3] and [4], and H. Theorem 13.2.2 were used. Then H. Theorem 13.2.1 and Wilson's theorems, i.e. Theorems 13.4.2 and 13.4.3, were applied a number of times. The result produced the interval $a = 5037 \leq n \leq b = 60000$ in which $N(n) \geq 7$. We now apply the corollary to Theorem 13.6.1. Since $11a + 16 < b$ we can replace b by any larger number! We have proved:

THEOREM 13.6.2. $n_7 \leq 5036$.

This is the best bound we could find using the methods of this chapter. Everything could be improved if one could prove that $N(12)$ is at least 7.

For the sake of completeness we point out that Wilson ([7]) improved the inequality of H. Theorem 13.4.2 to $N(v) \geq v^{1/17}$ for $v \geq v_0$.

References:

[1] H. Hanani, On the Number of Orthogonal Latin Squares, J. Comb. Theory 8 (1970), 247-271.

[2] D.R. Hughes and F.C. Piper, Projective Planes, Springer Verlag, New York 1973.

[3] R.C. Bose and S.S. Shrikhande, On the Construction of Sets of Mutually Orthogonal Latin Squares and the Falsity of a Conjecture of Euler, Trans. Am. Math. Soc. 95 (1960), 191-209.

[4] R.C. Bose, S.S. Shrikhande and E.T. Parker, Further Results on the Construction of Mutually Orthogonal Latin Squares and the Falsity of Euler's Conjecture, Can. J. of Math. 12 (1960), 189-203.

[5] A. Hedayat, A Set of Three Mutually Orthogonal Latin Squares of Order 15, Technometrics 13 (1971), 696-698.

[6] C.-C. Shih, A Method of Constructing Orthogonal Latin Squares (Chinese), Shuxhue Jinzhan 8 (1965), 98-104.

[7] R.M. Wilson, Concerning the Number of Mutually Orthogonal Latin Squares, Discrete Mathematics (submitted).

[8] G. Tarry, Le Problème des 36 Officiers, C.R. Assoc. Fr. Av. Sci. 1 (1900), 122-123 (1901), 170-203.

XIV. HADAMARD MATRICES

14.1. *More about* C- *and* S-*matrices.*

In § 10.1 we defined analogs of Hadamard matrices, called C-matrices and S-matrices (cf. 10.1.21, 10.1.22). We have already remarked that the matrix occurring in Paley's construction of Hadamard matrices (cf. H.14.1.23) is a skew C-matrix. The Kronecker product of matrices (cf. H. § 14.1) which turns out to be so useful for the construction of Hadamard matrices can be used in a similar fashion to produce statements concerning C-matrices. We consider symmetric or skew C-matrices of order v and write these in the form introduced in § 10.1:

$$C = \begin{pmatrix} 0 & \underline{j}^T \\ \pm\underline{j} & S \end{pmatrix},$$

where S is an S-matrix of order $v-1$. We present a number of theorems due to J.-M. Goethals and J.J. Seidel [1].

THEOREM 14.1.1. *If the pair* S_n, C_{n+1} *exists (symmetric or skew), then a pair* S_{n^2}, C_{n^2+1}, *both symmetric, exists.*

Proof. Take $S_{n^2} := S_n \times S_n + I_n \times J_n - J_n \times I_n$. The assertion is checked by straightforward calculation.

THEOREM 14.1.2. *There exists a symmetric* C-*matrix of order* $(n-1)^2+1$ *if* n *is of the form* $n = p^r + 1 \equiv 2 \pmod 4$ *or* $n = 2^t \prod_{i=1}^{s} (p_i^{r_i} + 1)$, $p_i^{r_i} + 1 \equiv 0 \pmod 4$, $i = 1,\dots,s$.

Proof. A symmetric pair S_{n-1}, C_n where $n = p^r + 1$ is given in (H.14.1.23), a skew pair of the order $2^t \prod_{i=1}^{s} (p_i^{r_i}+1)$ is given in H. Lemma 14.1.6. Hence the theorem follows from Theorem 14.1.1.

The following theorem extends H. Theorem 14.3.1.

THEOREM 14.1.3. *If symmetric or skew* C-*matrices of orders* n *and* $n+2$ *exist, then a Hadamard matrix of order* n^2 *exists.*

Proof. If the pairs S_{n-1}, C_n and S_{n+1}, C_{n+2} exist, then one pair is symmetric and the other is skew (cf. H.p.205). Now take

$$K := S_{n-1} \times S_{n+1} + I_{n-1} \times J_{n+1} - J_{n-1} \times I_{n+1} - I_{n-1} \times I_{n+1}.$$

Then K has order n^2-1, elements ± 1, and K satisfies the equations $KK^T = n^2 I - J$, $KJ = JK = J$. Hence

$$H := \begin{pmatrix} -1 & \underline{j}^T \\ \underline{j} & K \end{pmatrix}$$

is a Hadamard matrix of order n^2.

THEOREM 14.1.4. *If a Hadamard matrix of order* $m > 1$ *and a symmetric C-matrix of order* n *exist, then a Hadamard matrix of order* mn *exists.*

Proof. Let

$$P_m := \begin{pmatrix} 0 & -1 \\ 1 & 0 \end{pmatrix} \times I_{\frac{1}{2}m} .$$

If H_m is a Hadamard matrix and C_n a symmetric C-matrix, then

$$H_{mn} = H_m \times C_n + P_m H_m \times I_n$$

is a Hadamard matrix of order mn (straightforward calculation).

Corollary. If n has the form of Theorem 14.1.2 and if a Hadamard matrix of order $m > 1$ exists, then a Hadamard matrix of order $m((n-1)^2 + 1)$ exists.

Proof. This follows from Theorem 14.1.2.

Remark. Theorem 14.1.4 is of the same type as H. Lemma 14.1.5.

THEOREM 14.1.5. *If a Hadamard matrix of order* $m > 1$ *and a symmetric C-matrix of order* n *exist, then a Hadamard matrix of order* $mn(n-1)$ *exists.*

Proof. Take P_m as in the proof of Theorem 14.1.4. Let H_m be a Hadamard matrix; S_{n-1} and C_n a symmetric pair. Then

$$K = H_m \times C_n \times S_{n-1} + P_m H_m \times C_n \times I_{n-1} + H_m \times I_n \times J_{n-1}$$

is a Hadamard matrix of order $mn(n-1)$.

Remark. This theorem generalizes H. Theorem 14.1.3 to the symmetric case. A similar generalization of H. Theorem 14.1.4 is given in [1].

Furthermore, we mention without proof the following interesting theorem of R.J. Turyn [2] which generalizes Theorem 14.1.1.

THEOREM 14.1.6. *If there is a C-matrix (symmetric or skew) of order* $m+1$*, then there is a C-matrix of order* $m^n + 1$ *for every integer* n.

14.2. *A recent theorem of R.J. Turyn.*

At the end of H. § 14.2 it is remarked that no infinite class of Williamson-type Hadamard matrices has been found although it has been conjectured that they exist for all orders $n \equiv 0 \pmod 4$. Recently, R.J. Turyn [3] constructed such an infinite family. The construction depends on a special form of the Paley matrix (H.14.1.23). It was conjectured by V. Belevitch that this form is possible and this was proved by J.-M. Goethals and J.J. Seidel [1]. We present a proof along the lines suggested in [3].

We consider the construction of Paley matrices as described in H. § 14.1. Let χ be the quadratic character on GF(q), where $q \equiv 1 \pmod 4$. The elements of GF(q) are numbered $a_1,\dots,a_q$ and the matrix $Q := [q_{ij}]$ is defined by $q_{ij} := \chi(a_i - a_j)$.

We consider the Paley matrix

(14.2.1) $$P := \begin{pmatrix} 0 & \underline{j}^T \\ \underline{j} & Q \end{pmatrix} .$$

Now let R be a twodimensional vector space over GF(q). The pairwise linearly independent vectors

(14.2.2) $$\underline{x}_0 := \begin{pmatrix} 0 \\ 1 \end{pmatrix}, \quad \underline{x}_i := \begin{pmatrix} 1 \\ a_i \end{pmatrix}, \qquad (i = 1,2,\dots,q)$$

represent the lines through the origin in R. We can now write the matrix P of (14.2.1) as

(14.2.3) $$P = [p_{ij}] , \quad \text{where} \quad p_{ij} := \chi(\det(\underline{x}_i,\underline{x}_j)) .$$

Note that if we multiply any $\underline{x}_i$ by an element of GF(q), the vectors remain pairwise linearly independent and P is transformed into an equivalent matrix (row i and column i of P are multiplied by ± 1).

The field $GF(q^2)$ can be considered as the vector space R over GF(q). We take as standard basis $1 = \binom{1}{0}$ and $\xi = \binom{0}{1}$, where ξ is a primitive element of $GF(q^2)$. Then ξ satisfies an irreducible equation $\xi^2 + \gamma\xi + \alpha = 0$, where $\alpha = \xi^{q+1}$ is a primitive element of GF(q). We observe that multiplication by ξ in $GF(q^2)$ is a linear transformation τ of R with matrix $\begin{pmatrix} 0 & -\alpha \\ 1 & -\gamma \end{pmatrix}$, i.e. with determinant α. We shall use this in the proof of the following theorem.

THEOREM 14.2.1. *Let* $q \equiv 1 \pmod 4$. *The matrix* P *of* (14.2.1) *is equivalent to a matrix which has the form*

$$\hat{P} = \begin{pmatrix} U & V \\ V & -U \end{pmatrix} ,$$

where U *and* V *are symmetric circulants of order* $\frac{q+1}{2}$.

Proof. Write $q+1 = 2t$ (i.e. t is odd). We replace the vectors of (14.2.2) by the vectors

$$\underline{x}_i := \begin{cases} \xi^{4i} & \text{if } 0 \le i \le t-1 , \\ \xi^{4(i-t)+t} & \text{if } t \le i \le 2t-1 . \end{cases}$$

It is simple to check that the $\underline{x}_i$ are pairwise linearly independent. We have

$$\chi(\det(\xi^{4i},\xi^{4t})) = \chi(\det(\xi^{4i},\alpha^2)) = \chi(\det(\xi^{4i},\xi^0)) ,$$

i.e. U, and hence also V, is a circulant. $\hat{P}$ is clearly symmetric. We have, since α is not a square in GF(q),

$$v_{ij} = \chi(\det(\xi^{4i},\xi^{4j+t})) = \chi((\det \tau)^{-t} \det(\xi^{4i+t},\alpha\xi^{4j})) =$$

$$= - \chi(\det(\xi^{4i+t},\alpha\xi^{4j})) = \chi(\det(\xi^{4i+t},\xi^{4j})) .$$

This proves that V is symmetric. Finally,

$$\chi(\det(\xi^{4i+t},\xi^{4j+t})) = \chi((\det \tau)^{t} \det(\xi^{4i},\xi^{4j})) = - u_{ij} ,$$

proving that the lower right-hand side of $\hat{P}$ is $-U$.

We consider the following four symmetric circulants of size $\frac{q+1}{2}$ with entries ± 1:

(14.2.4) $\qquad A := I + U , \quad B := - I + U , \quad C := D := V .$

Since A, B, C and D are circulants they commute with each other. Furthermore,

$$A^2 + B^2 + C^2 + D^2 = (2q+2)I_{\frac{q+1}{2}}$$

because $\hat{P}^2 = qI_{q+1}$. We can therefore apply the result of Baumert and Hall (cf. H. § 14.3), thus proving

THEOREM 14.2.2. *If* q *is a prime power* $\equiv 1 \pmod 4$ *then there exists a Hadamard matrix of order* $6(q+1)$.

14.3. *Application of the construction methods.*

We consider the list of construction methods I to X on H.p.207 for orders $N \le 1000$. There are 250 multiples of 4 to be considered. We found

(a) 94 orders for which methods I and II yield Hadamard matrices.

(b) 102 orders for which method III yields a Hadamard matrix not found in (a). In the range we covered 520 is the only order where method III essentially uses an order h from II.

(c) Method IV produces 756 as only new order.

(d) Method V clearly yields the new orders 92, 116, 172.

(e) Methods VI to VIII produce no new results.

(f) From method IX the new order 324 is found.

(g) Method X then gives us the new orders 184 = 2.92, 232 = 2.116.

We can fill a number of the gaps which are left as follows:

(h) The order 156 by the Baumert-Hall method (H. § 14.3).

(i) The order 452 by applying the corollary to Theorem 14.1.4 with $m = 2$, $n = 16$. The order 904 from X.

(j) The orders 372, 612 and 732 as applications of Theorem 14.2.2 to the primes 61, 101 and the prime power 11^2.

(k) A number of recent methods of R.J. Turyn, J. Cooper and J. Wallis (cf. [4]) yields the orders 260, 476, 532, 836 and 988.

For the following orders < 1000 Hadamard matrices are unknown to us: 188, 236, 268, 292, 356, 376, 404, 412, 428, 436, 472, 508, 536, 584, 596, 604, 652, 668, 712, 716, 764, 772, 808, 852, 856, 872, 876, 892, 932, 940, 944, 952, 956, 964, 980, 996.

From this report on our activities we omit a number of results on symmetric and skew Hadamard matrices since these would overlap with a fairly complete survey by J.S. Wallis which appeared recently. We refer the interested reader to Part 4 of Springer Lecture Notes 292 ([4]).

References:

[1] J.-M. Goethals and J.J. Seidel, Orthogonal Matrices with Zero Diagonal, Can. J. Math. 19 (1967), 1001-1010.

[2] R.J. Turyn, On C-Matrices of Arbitrary Powers, Can. J. Math. 23 (1971), 531-535.

[3] R.J. Turyn, An Infinite Class of Williamson Matrices, J. Comb. Theory (A) 12 (1972), 319-321.

[4] W.D. Wallis, A.P. Street and J.S. Wallis, Combinatorics: Room Squares, Sum Free Sets, Hadamard Matrices, Lecture Notes in Mathematics 292, Springer Verlag, Berlin (1972).

XV. CONSTRUCTIONS OF BLOCK DESIGNS

15.1. *Repeated blocks.*

In § 10.2 we discussed primitive repetition designs (PRD_i) and we gave an example of a simple construction of a PRD_1. A natural question to ask is whether the methods of H. Chapter 15 can be used or generalized to construct primitive repetition designs. Hanani's methods (H. § 15.4) often yield designs with repeated blocks. In fact we shall consider in § 15.2 the question whether it is possible to avoid this. In this section we give a few examples. For more information concerning the construction of PRD's we refer the reader to [1].

Example 15.1.1. To construct a BD(10,3;30,9,2) using H. Lemma 15.4.4 one starts with a design B[{3,4,6},1,10]. This is obtained by taking EG(2,3) and adding a point to 3 blocks of one parallel class. We then have 9 blocks of 3 elements and 3 blocks of 4 elements. For these we substitute $\binom{1\ 1\ 1}{1\ 1\ 1}$, respectively $J_4 - I_4$. This yields the required design which then has 9 repeated blocks (each occurring twice). This is a PRD_2.

Example 15.1.2. We generalize (H.15.3.18). Let $v = 12t + 4$ and let A be the cyclic group of order v. The base blocks

$$[0\,,\,3t-i\,,\,3t+2+i]\quad,\quad i = 0,1,\ldots,t-1\ ,\quad \text{all twice}\ ,$$

$$[0\,,\,5t+1-i\,,\,5t+2+i]\ ,\quad i = 0,1,\ldots,t-1\ ,\quad \text{all twice}\ ,$$

$$[0\,,\,3t+1\,,\,6t+2]$$

form a base for a $BD(12t+4\,,\,3\,;\,(4t+1)(12t+4)\,,\,12t+3\,,\,2)$ in which $2t(12t+4)$ blocks are repeated twice.

Proof. The first two series of base blocks yield all differences $\pm\, j$ ($j = 1,\ldots,6t+1$) except for the difference $3t+1$, each occurring twice. The final block yields $\pm\,(3t+1)$ twice and $6t+2$ twice, because $6t+2 = -(6t+2) \pmod v$. Taking $t = 1$ we find a PRD_{15}.

Example 15.1.3. We now generalize H. Theorem 15.3.4. Let $v = 6t+1 = p^n$, p a prime, $t \geq 2$. Let x be a primitive element in the field $GF(p^n)$ and $x^{2t} - 1 = x^s$. Then the blocks

$$[x^i, x^{2t+i}, x^{4t+i}]\ ,\quad \text{taken five times for } i = 0 \text{ and six times for } i = 1,2,\ldots,t-1\ ,$$

$$[\infty,0,x^s]\ ,\ [\infty,0,x^{s+t}]\ ,\ [\infty,0,x^{s+2t}]$$

are base blocks with respect to the additive group of $GF(p^n)$ of a

BD(6t + 2 , 3 ; (6t + 1)(6t + 2) , 3(6t + 1) , 6) with certain blocks repeated 6 times.

Proof. The differences $x^{s+i+\ell t}$, $\ell = 0,1,\dots,5$; $i = 1,2,\dots,t-1$, occur 6 times. If we take $i = 0$ the difference $x^{s+i+\ell t}$ occurs 5 times among the repeated base blocks and each such difference occurs once in the last 3 base blocks. These also yield all mixed differences with ∞. If we take $t = 2$ we find a PRD_9.

15.2. *Steiner triple systems.*

We make several observations concerning Moore's construction of STS's, i.e. H. Theorem 15.4.2. First we consider the construction a little more closely. Let there be a STS of order v_2 containing a subsystem of order v_3 (or $v_3 = 1$) and let there be a STS of order v_1. As in (H. 15.4.8) we define $S_0 := \{a_1,\dots,a_{v_3}\}$, $S_i := \{b_{i1},\dots,b_{is}\}$ for $i = 1,2,\dots,v_1$, $s = v_2 - v_3 = 2\ell$. On S_0 there is a STS (step 1 on H.p.218). On each of the $S_0 \cup S_i$ we have a STS (step 2 on H.p.239), containing the STS on S_0. The particular way in which we number the points of S_i is irrelevant. Take any block of the design on v_1 points, say (j,k,r). According to step 3 of the construction we have among others the blocks

$$(b_{j,1},b_{k,\ell+1},b_{r,\ell-2}) , \quad (b_{j,\ell+1},b_{k,\ell+1},b_{r,2\ell-2})$$

$$(b_{j,1},b_{k,1},b_{r,2\ell-2}) , \quad (b_{j,\ell+1},b_{k,1},b_{r,\ell-2}) .$$

We now see to it that the construction in step 2 has

$(a_1,b_{j,1},b_{j,\ell+1})$ as a block of the STS on $S_0 \cup S_j$,

$(a_1,b_{k,1},b_{k,\ell+1})$ as a block of the STS on $S_0 \cup S_k$,

$(a_1,b_{r,\ell-2},b_{r,2\ell-2})$ as a block of the STS on $S_0 \cup S_r$.

We have followed completely the construction of H. Theorem 15.4.2 and we have proved:

THEOREM 15.2.1. *If there is a* STS *of order* v_2 *containing a subsystem of order* v_3 *(or* $v_3 = 1$*) and if there is a* STS *of order* v_1 *then there is a* STS *of order* $v = v_3 + v_1(v_2 - v_3)$ *which contains* v_1 *subsystems of order* v_2*, one of order* v_1*, one of order* v_3 *and furthermore one of order 7.*

Examples. Since $25 = 1 + 3(9-1)$, $27 = 1 + 13(3-1)$, $33 = 3 + 3(13-3)$, $37 = 1 + 3(13-1)$ we can construct STS's of these orders containing PG(2,2) as a subsystem. If we now use Table 15.1 on H.p.240 we can immediately show that there is a STS of order v if $v = 6t + 1$ or $v = 6t + 3$ (H. Theorem 15.4.3). This is what is suggested at the bottom of H.p.241 and quite a lot easier than the proof as presented.

A different simplification of Moore's construction, avoiding subsystems of order 7 completely, was given recently by A.J.W. Hilton [2].
Moore's construction can be generalized to other values of k. We mention one such generalization given by R.M. Wilson in [3], Part I. He attributes it to D.K. Ray-Chaudhuri.

THEOREM 15.2.2. *If there is a block design with* $\lambda = 1$ *on* $m + d$ *points, with blocks of size* k, *containing a subdesign on* d *points, and if there is an* OA(m,k), *then for every* $v \in B(k,1)$ *we have* $mv + d \in B(k,1)$.

The special case $m = v_2 - v_3$, $d = v_3$, $v = v_1$, $k = 3$ is Moore's theorem. We mention that Theorem 11.6 in [3], Part I also generalizes Moore's construction.

We have already observed that Hanani's constructions of block designs with k = 3 often lead to designs with repeated blocks. One way of avoiding this is copying the constructions using sets K_i^j which do not contain 3 (cf. H. Lemma 15.4.2, for instance). We have done this but it does not seem interesting enough to include in this report. The first case one would be led to try is $\lambda = 2$. Note that in H. Theorem 15.4.4 the construction yields repeated blocks if $v = 6t + 1$ or $6t + 3$. This can be avoided by applying the following interesting result of J. Doyen [4]. Let D(v) denote the maximal number of STS's on v points which are pairwise disjoint.

THEOREM 15.2.3. *If* $v = 6t + 3$, *then*

$$D(v) \geq 4t + 1 \quad \text{if} \quad 2t + 1 \not\equiv 0 \pmod 3 ,$$

$$D(v) \geq 4t - 1 \quad \text{if} \quad 2t + 1 \equiv 0 \pmod 3 ;$$

if $v = 6t + 1$, *then*

$$D(v) \geq t/2 \quad \text{if} \quad t \equiv 0 \pmod 2 ,$$

$$D(v) \geq 2t - 1 \quad \text{if} \quad t \equiv 1 \pmod 2 .$$

In his paper Doyen observes (Corollary 2 on p.412) that this implies that a triple system with $\lambda = 2$ and $v = 6t + 1$ or $6t + 3$ without repeated blocks is possible. Since it is easily seen that the constructions of (H.15.3.17) and (H.15.3.18) never give repeated blocks, this completely settles the question for all possible $v \neq 3$ and $\lambda = 2$. In the same way one could attack the case $\lambda = 3$. Obviously Theorem 15.2.3 covers $v \equiv 1$ or 3 (mod 6) except for v = 3, 7, 13, 25. Clearly, repetition of blocks is necessary for v = 3. If v = 7 we can take a triple system with $\lambda = 2$. The complement (i.e. all other blocks of size 3) is then a triple system with $\lambda = 3$ and no repeated blocks. For v = 13 or 25 a special construction is necessary (we omit this). It remains to consider $v \equiv 5$ (mod 6). Here we refer to a recent paper by C.St.J.A. Nash-Williams

[5]. Although he does not exclude repeated blocks, his method has the property that for $v \equiv 5 \pmod 6$ there are no repeated blocks. If one wishes to pursue this question it is clear that λ must be restricted. Obviously, if $\lambda > v-2$, then repetition of blocks is unavoidable. For $v = 6t + 3$ we have at least $4t - 1$ disjoint STS's by Theorem 15.2.3. By combining we cover all cases $\lambda \le 4t - 1$ and by taking complements we settle the remaining cases. However, other values of v are not settled. This seems a worthwhile problem to analyse.

15.3. *Recent results.*

To finish this chapter and also this report we point out 3 recent papers which are connected with H. Chapter 15.

(a) The "Kirkman schoolgirl problem" mentioned on H.p.241, was settled by D.K. Ray-Chaudhuri and R.M. Wilson [6].

(b) The existence conjecture for block designs (H.p.248) was settled for $k = 5$ (cf. H.p.250, line 1) by H. Hanani [7]. There is one exception, i.e. $v = 15$, $b = 21$, $r = 7$, $k = 5$, $\lambda = 2$.

(c) In [3] R.M. Wilson proves the existence conjecture with the condition $k/(k,\lambda) = 1$ or p^{α}, or $\lambda \ge ([\frac{1}{2}k] - 1)([\frac{1}{2}k] - 2)$.

(d) In [8] R.M. Wilson proves the general existence conjecture.

References:

[1] J.H. van Lint, Block Designs with Repeated Blocks and $(b,r,\lambda) = 1$, J. Comb. Theory (to appear).

[2] A.J.W. Hilton, A Simplification of Moore's Proof of the Existence of Steiner Triple Systems, J. Comb. Theory (A) 13 (1972), 422-425.

[3] R.M. Wilson, An Existence Theory for Pairwise Balanced Designs, I: Composition Theorems and Morphisms, II: The Structure of PBD-Closed Sets and the Existence Conjectures, J. Comb. Theory (A) 13 (1972), 220-245 and 246-273.

[4] J. Doyen, Constructions of Disjoint Steiner Triple Systems, Proc. A.M.S. 32 (1972), 409-416.

[5] C.St.J.A. Nash-Williams, Simple Constructions for Balanced Incomplete Block Designs with Block Size Three, J. Comb. Theory (A) 13 (1972), 1-6.

[6] D.K. Ray-Chaudhuri and R.M. Wilson, Solution of Kirkman's Schoolgirl Problem, "Combinatorics", A.M.S. Proc. Symp. Pure Math. 9 (1971), 187-203.

[7] H. Hanani, On Balanced Incomplete Block Designs with Blocks Having Five Elements, J. Comb. Theory (A) 12 (1972), 184-201.

[8] R.M. Wilson, An Existence Theory for Pairwise Balanced Designs, III: Proof of the Existence Conjectures (submitted).

INDEX

Vol. 215: P. Antonelli, D. Burghelea and P. J. Kahn, The Concordance-Homotopy Groups of Geometric Automorphism Groups. X, 140 pages. 1971. DM 16,-

Vol. 216: H. Maaß, Siegel's Modular Forms and Dirichlet Series. VII, 328 pages. 1971. DM 20,-

Vol. 217: T. J. Jech, Lectures in Set Theory with Particular Emphasis on the Method of Forcing. V, 137 pages. 1971. DM 16,-

Vol. 218: C. P. Schnorr, Zufälligkeit und Wahrscheinlichkeit. IV, 212 Seiten. 1971. DM 20,-

Vol. 219: N. L. Alling and N. Greenleaf, Foundations of the Theory of Klein Surfaces. IX, 117 pages. 1971. DM 16,-

Vol. 220: W. A. Coppel, Disconjugacy. V, 148 pages. 1971. DM 16,-

Vol. 221: P. Gabriel und F. Ulmer, Lokal präsentierbare Kategorien. V, 200 Seiten. 1971. DM 18,-

Vol. 222: C. Meghea, Compactification des Espaces Harmoniques. III, 108 pages. 1971. DM 16,-

Vol. 223: U. Felgner, Models of ZF-Set Theory. VI, 173 pages. 1971. DM 16,-

Vol. 224: Revètements Etales et Groupe Fondamental. (SGA 1). Dirigé par A. Grothendieck XXII, 447 pages. 1971. DM 30,-

Vol. 225: Théorie des Intersections et Théorème de Riemann-Roch. (SGA 6). Dirigé par P. Berthelot, A. Grothendieck et L. Illusie. XII, 700 pages. 1971. DM 40,-

Vol. 226: Seminar on Potential Theory, II. Edited by H. Bauer. IV, 170 pages. 1971. DM 18,-

Vol. 227: H. L. Montgomery, Topics in Multiplicative Number Theory. IX, 178 pages. 1971. DM 18,-

Vol. 228: Conference on Applications of Numerical Analysis. Edited by J. Ll. Morris. X, 358 pages. 1971. DM 26,-

Vol. 229: J. Väisälä, Lectures on n-Dimensional Quasiconformal Mappings. XIV, 144 pages. 1971. DM 16,-

Vol. 230: L. Waelbroeck, Topological Vector Spaces and Algebras. VII, 158 pages. 1971. DM 16,-

Vol. 231: H. Reiter, L^1-Algebras and Segal Algebras. XI, 113 pages. 1971. DM 16,-

Vol. 232: T. H. Ganelius, Tauberian Remainder Theorems. VI, 75 pages. 1971. DM 16,-

Vol. 233: C. P. Tsokos and W. J. Padgett. Random Integral Equations with Applications to stochastic Systems. VII, 174 pages. 1971. DM 18,-

Vol. 234: A. Andreotti and W. Stoll. Analytic and Algebraic Dependence of Meromorphic Functions. III, 390 pages. 1971. DM 26,-

Vol. 235: Global Differentiable Dynamics. Edited by O. Hájek, A. J. Lohwater, and R. McCann. X, 140 pages. 1971. DM 16,-

Vol. 236: M. Barr, P. A. Grillet, and D. H. van Osdol. Exact Categories and Categories of Sheaves. VII, 239 pages. 1971. DM 20,-

Vol. 237: B. Stenström, Rings and Modules of Quotients. VII, 136 pages. 1971. DM 16,-

Vol. 238: Der kanonische Modul eines Cohen-Macaulay-Rings. Herausgegeben von Jürgen Herzog und Ernst Kunz. VI, 103 Seiten. 1971. DM 16,-

Vol. 239: L. Illusie, Complexe Cotangent et Déformations I. XV, 355 pages. 1971. DM 26,-

Vol. 240: A. Kerber, Representations of Permutation Groups I. VII, 192 pages. 1971. DM 18,-

Vol. 241: S. Kaneyuki, Homogeneous Bounded Domains and Siegel Domains. V, 89 pages. 1971. DM 16,-

Vol. 242: R. R. Coifman et G. Weiss, Analyse Harmonique Non-Commutative sur Certains Espaces. V, 160 pages. 1971. DM 16,-

Vol. 243: Japan-United States Seminar on Ordinary Differential and Functional Equations. Edited by M. Urabe. VIII, 332 pages. 1971. DM 26,-

Vol. 244: Séminaire Bourbaki - vol. 1970/71. Exposés 382-399. IV, 356 pages. 1971. DM 26,-

Vol. 245: D. E. Cohen, Groups of Cohomological Dimension One. V, 99 pages. 1972. DM 16,-

Vol. 246: Lectures on Rings and Modules. Tulane University Ring and Operator Theory Year, 1970-1971. Volume I. X, 661 pages. 1972. DM 40,-

Vol. 247: Lectures on Operator Algebras. Tulane University Ring and Operator Theory Year, 1970-1971. Volume II. XI, 786 pages. 1972. DM 40,-

Vol. 248: Lectures on the Applications of Sheaves to Ring Theory. Tulane University Ring and Operator Theory Year, 1970-1971. Volume III. VIII, 315 pages. 1971. DM 26,-

Vol. 249: Symposium on Algebraic Topology. Edited by P. J. Hilton. VII, 111 pages. 1971. DM 16,-

Vol. 250: B. Jónsson, Topics in Universal Algebra. VI, 220 pages. 1972. DM 20,-

Vol. 251: The Theory of Arithmetic Functions. Edited by A. A. Gioia and D. L. Goldsmith VI, 287 pages. 1972. DM 24,-

Vol. 252: D. A. Stone, Stratified Polyhedra. IX, 193 pages. 1972. DM 18,-

Vol. 253: V. Komkov, Optimal Control Theory for the Damping of Vibrations of Simple Elastic Systems. V, 240 pages. 1972. DM 20,-

Vol. 254: C. U. Jensen, Les Foncteurs Dérivés de $\varprojlim$ et leurs Applications en Théorie des Modules. V, 103 pages. 1972. DM 16,-

Vol. 255: Conference in Mathematical Logic - London '70. Edited by W. Hodges. VIII, 351 pages. 1972. DM 26,-

Vol. 256: C. A. Berenstein and M. A. Dostal, Analytically Uniform Spaces and their Applications to Convolution Equations. VII, 130 pages. 1972. DM 16,-

Vol. 257: R. B. Holmes, A Course on Optimization and Best Approximation. VIII, 233 pages. 1972. DM 20,-

Vol. 258: Séminaire de Probabilités VI. Edited by P. A. Meyer. VI, 253 pages. 1972. DM 22,-

Vol. 259: N. Moulis, Structures de Fredholm sur les Variétés Hilbertiennes. V, 123 pages. 1972. DM 16,-

Vol. 260: R. Godement and H. Jacquet, Zeta Functions of Simple Algebras. IX, 188 pages. 1972. DM 18,-

Vol. 261: A. Guichardet, Symmetric Hilbert Spaces and Related Topics. V, 197 pages. 1972. DM 18,-

Vol. 262: H. G. Zimmer, Computational Problems, Methods, and Results in Algebraic Number Theory. V, 103 pages. 1972. DM 16,-

Vol. 263: T. Parthasarathy, Selection Theorems and their Applications. VII, 101 pages. 1972. DM 16,-

Vol. 264: W. Messing, The Crystals Associated to Barsotti-Tate Groups: With Applications to Abelian Schemes. III, 190 pages. 1972. DM 18,-

Vol. 265: N. Saavedra Rivano, Catégories Tannakiennes. II, 418 pages. 1972. DM 26,-

Vol. 266: Conference on Harmonic Analysis. Edited by D. Gulick and R. L. Lipsman. VI, 323 pages. 1972. DM 24,-

Vol. 267: Numerische Lösung nichtlinearer partieller Differential- und Integro-Differentialgleichungen. Herausgegeben von R. Ansorge und W. Törnig, VI, 339 Seiten. 1972. DM 26,-

Vol. 268: C. G. Simader, On Dirichlet's Boundary Value Problem. IV, 238 pages. 1972. DM 20,-

Vol. 269: Théorie des Topos et Cohomologie Etale des Schémas. (SGA 4). Dirigé par M. Artin, A. Grothendieck et J. L. Verdier. XIX, 525 pages. 1972. DM 50,-

Vol. 270: Théorie des Topos et Cohomologie Etale des Schémas. Tome 2. (SGA 4). Dirigé par M. Artin, A. Grothendieck et J. L. Verdier. V, 418 pages. 1972. DM 50,-

Vol. 271: J. P. May, The Geometry of Iterated Loop Spaces. IX, 175 pages. 1972. DM 18,-

Vol. 272: K. R. Parthasarathy and K. Schmidt, Positive Definite Kernels, Continuous Tensor Products, and Central Limit Theorems of Probability Theory. VI, 107 pages. 1972. DM 16,-

Vol. 273: U. Seip, Kompakt erzeugte Vektorräume und Analysis. IX, 119 Seiten. 1972. DM 16,-

Vol. 274: Toposes, Algebraic Geometry and Logic. Edited by. F. W. Lawvere. VI, 189 pages. 1972. DM 18,-

Vol. 275: Séminaire Pierre Lelong (Analyse) Année 1970-1971. VI, 181 pages. 1972. DM 18,-

Vol. 276: A. Borel, Représentations de Groupes Localement Compacts. V, 98 pages. 1972. DM 16,-

Vol. 277: Séminaire Banach. Edité par C. Houzel. VII, 229 pages. 1972. DM 20,-

Vol. 278: H. Jacquet, Automorphic Forms on GL(2). Part II. XIII, 142 pages. 1972. DM 16,–

Vol. 279: R. Bott, S. Gitler and I. M. James, Lectures on Algebraic and Differential Topology. V, 174 pages. 1972. DM 18,–

Vol. 280: Conference on the Theory of Ordinary and Partial Differential Equations. Edited by W. N. Everitt and B. D. Sleeman. XV, 367 pages. 1972. DM 26,–

Vol. 281: Coherence in Categories. Edited by S. Mac Lane. VII, 235 pages. 1972. DM 20,–

Vol. 282: W. Klingenberg und P. Flaschel, Riemannsche Hilbertmannigfaltigkeiten. Periodische Geodätische. VII, 211 Seiten. 1972. DM 20,–

Vol. 283: L. Illusie, Complexe Cotangent et Déformations II. VII, 304 pages. 1972. DM 24,–

Vol. 284: P. A. Meyer, Martingales and Stochastic Integrals I. VI, 89 pages. 1972. DM 16,–

Vol. 285: P. de la Harpe, Classical Banach-Lie Algebras and Banach-Lie Groups of Operators in Hilbert Space. III, 160 pages. 1972. DM 16,–

Vol. 286: S. Murakami, On Automorphisms of Siegel Domains. V, 95 pages. 1972. DM 16,–

Vol. 287: Hyperfunctions and Pseudo-Differential Equations. Edited by H. Komatsu. VII, 529 pages. 1973. DM 36,–

Vol. 288: Groupes de Monodromie en Géométrie Algébrique. (SGA 7 I). Dirigé par A. Grothendieck. IX, 523 pages. 1972. DM 50,–

Vol. 289: B. Fuglede, Finely Harmonic Functions. III, 188. 1972. DM 18,–

Vol. 290: D. B. Zagier, Equivariant Pontrjagin Classes and Applications to Orbit Spaces. IX, 130 pages. 1972. DM 16,–

Vol. 291: P. Orlik, Seifert Manifolds. VIII, 155 pages. 1972. DM 16,–

Vol. 292: W. D. Wallis, A. P. Street and J. S. Wallis, Combinatorics: Room Squares, Sum-Free Sets, Hadamard Matrices. V, 508 pages. 1972. DM 50,–

Vol. 293: R. A. DeVore, The Approximation of Continuous Functions by Positive Linear Operators. VIII, 289 pages. 1972. DM 24,–

Vol. 294: Stability of Stochastic Dynamical Systems. Edited by R. F. Curtain. IX, 332 pages. 1972. DM 26,–

Vol. 295: C. Dellacherie, Ensembles Analytiques, Capacités, Mesures de Hausdorff. XII, 123 pages. 1972. DM 16,–

Vol. 296: Probability and Information Theory II. Edited by M. Behara, K. Krickeberg and J. Wolfowitz. V, 223 pages. 1973. DM 20,–

Vol. 297: J. Garnett, Analytic Capacity and Measure. IV, 138 pages. 1972. DM 16,–

Vol. 298: Proceedings of the Second Conference on Compact Transformation Groups. Part 1. XIII, 453 pages. 1972. DM 32,–

Vol. 299: Proceedings of the Second Conference on Compact Transformation Groups. Part 2. XIV, 327 pages. 1972. DM 26,–

Vol. 300: P. Eymard, Moyennes Invariantes et Représentations Unitaires. II. 113 pages. 1972. DM 16,–

Vol. 301: F. Pittnauer, Vorlesungen über asymptotische Reihen. VI, 186 Seiten. 1972. DM 18,–

Vol. 302: M. Demazure, Lectures on p-Divisible Groups. V, 98 pages. 1972. DM 16,–

Vol. 303: Graph Theory and Applications. Edited by Y. Alavi, D. R. Lick and A. T. White. IX, 329 pages. 1972. DM 26,–

Vol. 304: A. K. Bousfield and D. M. Kan, Homotopy Limits, Completions and Localizations. V, 348 pages. 1972. DM 26,–

Vol. 305: Théorie des Topos et Cohomologie Etale des Schémas. Tome 3. (SGA 4). Dirigé par M. Artin, A. Grothendieck et J. L. Verdier. VI, 640 pages. 1973. DM 50,–

Vol. 306: H. Luckhardt, Extensional Gödel Functional Interpretation. VI, 161 pages. 1973. DM 18,–

Vol. 307: J. L. Bretagnolle, S. D. Chatterji et P.-A. Meyer, Ecole d'été de Probabilités: Processus Stochastiques. VI, 198 pages. 1973. DM 20,–

Vol. 308: D. Knutson, λ-Rings and the Representation Theory of the Symmetric Group. IV, 203 pages. 1973. DM 20,–

Vol. 309: D. H. Sattinger, Topics in Stability and Bifurcation Theory. VI, 190 pages. 1973. DM 18,–

Vol. 310: B. Iversen, Generic Local Structure of the Morphisms in Commutative Algebra. IV, 108 pages. 1973. DM 16,–

Vol. 311: Conference on Commutative Algebra. Edited by J. W. Brewer and E. A. Rutter. VII, 251 pages. 1973. DM 22,–

Vol. 312: Symposium on Ordinary Differential Equations. Edited by W. A. Harris, Jr. and Y. Sibuya. VIII, 204 pages. 1973. DM 22,–

Vol. 313: K. Jörgens and J. Weidmann, Spectral Properties of Hamiltonian Operators. III, 140 pages. 1973. DM 16,–

Vol. 314: M. Deuring, Lectures on the Theory of Algebraic Functions of One Variable. VI, 151 pages. 1973. DM 16,–

Vol. 315: K. Bichteler, Integration Theory (with Special Attention to Vector Measures). VI, 357 pages. 1973. DM 26,–

Vol. 316: Symposium on Non-Well-Posed Problems and Logarithmic Convexity. Edited by R. J. Knops. V, 176 pages. 1973. DM 18,–

Vol. 317: Séminaire Bourbaki – vol. 1971/72. Exposés 400–417. IV, 361 pages. 1973. DM 26,–

Vol. 318: Recent Advances in Topological Dynamics. Edited by A. Beck, VIII, 285 pages. 1973. DM 24,–

Vol. 319: Conference on Group Theory. Edited by R. W. Gatterdam and K. W. Weston. V, 188 pages. 1973. DM 18,–

Vol. 320: Modular Functions of One Variable I. Edited by W. Kuyk. V, 195 pages. 1973. DM 18,–

Vol. 321: Séminaire de Probabilités VII. Edité par P. A. Meyer. VI, 322 pages. 1973. DM 26,–

Vol. 322: Nonlinear Problems in the Physical Sciences and Biology. Edited by I. Stakgold, D. D. Joseph and D. H. Sattinger. VIII, 357 pages. 1973. DM 26,–

Vol. 323: J. L. Lions, Perturbations Singulières dans les Problèmes aux Limites et en Contrôle Optimal. XII, 645 pages. 1973. DM 42,–

Vol. 324: K. Kreith, Oscillation Theory. VI, 109 pages. 1973. DM 16,–

Vol. 325: Ch.-Ch. Chou, La Transformation de Fourier Complexe et L'Equation de Convolution. IX, 137 pages. 1973. DM 16,–

Vol. 326: A. Robert, Elliptic Curves. VIII, 264 pages. 1973. DM 22,–

Vol. 327: E. Matlis, 1-Dimensional Cohen-Macaulay Rings. XII, 157 pages. 1973. DM 18,–

Vol. 328: J. R. Büchi and D. Siefkes, The Monadic Second Order Theory of All Countable Ordinals. VI, 217 pages. 1973. DM 20,–

Vol. 329: W. Trebels, Multipliers for (C, α)-Bounded Fourier Expansions in Banach Spaces and Approximation Theory. VII, 103 pages. 1973. DM 16,–

Vol. 330: Proceedings of the Second Japan-USSR Symposium on Probability Theory. Edited by G. Maruyama and Yu. V. Prokhorov. VI, 550 pages. 1973. DM 36,–

Vol. 331: Summer School on Topological Vector Spaces. Edited by L. Waelbroeck. VI, 226 pages. 1973. DM 20,–

Vol. 332: Séminaire Pierre Lelong (Analyse) Année 1971-1972. V, 131 pages. 1973. DM 16,–

Vol. 333: Numerische, insbesondere approximationstheoretische Behandlung von Funktionalgleichungen. Herausgegeben von R. Ansorge und W. Törnig. VI, 296 Seiten. 1973. DM 24,–

Vol. 334: F. Schweiger, The Metrical Theory of Jacobi-Perron Algorithm. V, 111 pages. 1973. DM 16,–

Vol. 335: H. Huck, R. Roitzsch, U. Simon, W. Vortisch, R. Walden, B. Wegner und W. Wendland, Beweismethoden der Differentialgeometrie im Großen. IX, 159 Seiten. 1973. DM 18,–

Vol. 336: L'Analyse Harmonique dans le Domaine Complexe. Edité par E. J. Akutowicz. VIII, 169 pages. 1973. DM 18,–

Vol. 337: Cambridge Summer School in Mathematical Logic. Edited by A. R. D. Mathias and H. Rogers. IX, 660 pages. 1973. DM 42,–

Vol. 338: J. Lindenstrauss and L. Tzafriri, Classical Banach Spaces. IX, 243 pages. 1973. DM 22,–

Vol. 339: G. Kempf, F. Knudsen, D. Mumford and B. Saint-Donat, Toroidal Embeddings I. VIII, 209 pages. 1973. DM 20,–

Vol. 340: Groupes de Monodromie en Géométrie Algébrique. (SGA 7 II). Par P. Deligne et N. Katz. X, 438 pages. 1973. DM 40,–

Vol. 341: Algebraic K-Theory I, Higher K-Theories. Edited by H. Bass. XV, 335 pages. 1973. DM 26,–

Vol. 342: Algebraic K-Theory II, "Classical" Algebraic K-Theory, and Connections with Arithmetic. Edited by H. Bass. XV, 527 pages. 1973. DM 36,–